Telecommunications Demand: A Survey and Critique

Telecommunications Demand: A Survey and Critique

Lester D. Taylor
University of Arizona

Ballinger Publishing Company • Cambridge, Massachusetts
A Subsidiary of Harper & Row, Publishers, Inc.

International Standard Book Number: 0-88410-496-6

Library of Congress Catalog Card Number: 79-24836

Printed in the United States of America

Library of Congress Cataloging in Publication Data

Taylor, Lester D.
Telecommunications demand.

Bibliography: p.
Includes index.
1. Telephone—United States. 2. Telephone—Mathematical models.
I. Title.
HE8815.T39 384.6'0973 79-24836
ISBN 0-88410-496-6

Dedication

With love,
to Carol, Henk, and Becca.

Contents

List of Figures

List of Tables

Preface

This monograph provides a comprehensive review of econometric demand modeling in the telecommunications industry and has five main purposes: (1) to survey and critique existing research, (2) to make a start on developing a theoretical framework for analyzing telecommunications demand, (3) to bring together in one place the existing estimates of price and income elasticities of telecommunications demand, (4) to make recommendations for future research, and (5), above all, to make the literature on telecommunications demand available to a wider audience.

The study has been financed by the Regulatory Research Section, State Regulatory Matters Department, of the American Telephone and Telegraph Company. The manuscript has been reviewed extensively by persons both inside and outside the Bell System, but the conclusions expressed are mine and reflect my own interpretation of the theoretical and empirical literature and my own views concerning the telecommunications industry. In the course of the survey, I have reviewed a large number of Bell System demand studies, many of which are not yet (or may never be) in the public domain, and, to protect the proprietary interests of the Bell System operating telephone companies, it has been necessary to identify the models presented in Chapter 4 on intrastate toll demand by code. However, except for this instance, models and results are attributed to the individuals and organizations that spawned them, and I am grateful to the Bell System (and also to Bell Canada) for the willingness to allow numerous unpublished documents to be made known to a general audience.

The review is intended for a wide range of readers, and, except for Chapter 2, I have tried to keep the discussion nontechnical. Chapter 2, which is on the theoretical foundations of telecommunications demand, is primarily directed at my fellow economists, both inside and outside the telecommunications industry, who are interested in the more analytical aspects of telecommunications demand. Those who are only interested in an overview and conclusions can skip over Chapters 2–4 and just read Chapters 1 and 5. And those who are only interested in conclusions can restrict their attention to Chapter 5.

The literature focused on in this review is almost exclusively the econometric literature on telecommunications demand. Accordingly, the studies reviewed are based on time-series or cross-sectional samples that contain a sufficient number of data points for standard econometric and statistical techniques to be meaningful. Missing, therefore, is the broad range of forecasting models that are based on market surveys, interviews with potential customers, or Delphi techniques. Such forecasting models are of great importance in situations where the data base for econometric models does not exist or in the marketing of new products. My failure to review these techniques should not be interpreted as reflecting prejudice on my part against their usefulness. Time has not permitted me to be truly comprehensive in this critique, and, being an econometrician, it is natural that my focus has been on the econometric literature.

The assistance that has been provided by AT&T Regulatory Research, of which Peter B. Linhart is director, has been extraordinary. Edward D. Lowry, manager, Regulatory Research, coordinated the study from its beginning. In doing this, he was assisted admirably by Donna Coughlin and Lois Ramhold, who conducted an initial search of the literature and made it available to me in a highly organized form. Donna Coughlin also organized the bibliography and checked all of the text references for accuracy, and Lois Ramhold collated the extensive comments that were received on the first draft. In addition to these people, I am also grateful to Al Opie, director of the Public Switched Network Section (PSN) at AT&T, and Robert Mathieson, manager (PSN), Demand Analysis Group, for their special efforts in providing the documentation of the state toll demand models. These people have lessened my task considerably.

The monograph has benefited considerably from the extensive and thoughtful comments that were received on the first draft. In particular, I am grateful to the following persons for comments, criticism, or general discussions: Roy Billinghurst, Malcolm Burnside, James Cosgrove, Donald Degazon, Fredric Harbus, Michael Kennedy,

Lily Lai, Peter Linhart, William Lockhart, Edward Lowry, Catherine Martin, Patrick Marfisi, Robert Mathieson, Walter Paczkowski, Paul Robinson, A.L. Shultes, Subhabrata Sen. Charles Stevenson, Timothy Stewart, Susan Stuntebeck, Aleta Styers, and Michael Wert of AT&T General Departments; Robert Auray, Samuel Campbell, Robert Craver, Jeffrey Feldman, George Fuller, Lester Hendrickson, Linda Kalver, Kenneth Kleefeld, Richard Kuo, Mark Lemler, Richard Saunders, Scott Stephan, and Glen Thompsen of AT&T Long Lines; and Belinda Brandon, Paul Brandon, and Carl Pavarini of Bell Laboratories.

Continuing the list, Edward Dinan of New England Telephone Company, Edward Milker of Mountain Bell, Thomas Giddings of Bell Canada, Robert Couch of Bell Canada and the Trans Canada Telephone System, Lewis Perl and William Shew of the National Economic Research Associates, William J. Baumol of New York and Princeton Universities, H.S. Houthakker and Dale W. Jorgenson, Both of Harvard University, Daniel McFadden of M.I.T., Arthur M. Okun of the Brookings Institution, Merton J. Peck of Yale University, and Daniel Alger, Arthur Denzau, and John T. Wenders, colleagues at the University of Arizona. In addition, I have much appreciated the support and implicit trust of Carl E. Horn, assistant vice president of State Regulatory Matters at AT&T. Finally, I am grateful to Dee Dillhoff and Nancy McKinney at the University of Arizona for secretarial assistance, and to Rita Glacken and Kathy O'Neill at AT&T for proofreading and clerical assistance.

The cutoff date for the studies that are reviewed was April 1978. The study was completed in November 1978, and began circulating within the Bell System in early January 1979. However, there has been a lot of research since the April 1978 cutoff date, and some additional studies have been brought to my attention. As a consequence, I have added a postscript, Chapter 6, which draws attention to a couple of these earlier omissions and, in general, tries to bring things up to the present.

Finally, I should mention that this version of the monograph differs from that which initially circulated within the Bell System only in the addition of the postscript and the removal from Chapter 5 of a formal agenda of suggested research projects to be undertaken by the Bell System. Except for minor editing and the changing of a number here and there, the two versions are otherwise identical.

Wilson, Wyoming
August 1979

Lester D. Taylor

Chapter 1

Introduction and Overview

The focus in this study is on telephone demand. At the end of 1977, 162.1 million telephones were in operation in the United States, 119.8 million in residences, and 42.3 million in business and industry. Households spent $20.9 billion in 1976—nearly 2 percent of their total consumption expenditure—for telephone services. The American Telephone and Telegraph Company, which had revenues of $36.4 billion in 1977 and whose operatiing telephone companies service 79.3 percent of the telephones in the United States, is the largest nonfinancial corporation in the world. Thus telephone demand is big business, and one should expect a literature to match.

The literature on telephone demand is indeed large, but it is like an iceberg: much of it consists of unpublished memoranda, or else is contained in testimony before the FCC or state Public Utility Commissions, and is therefore not readily discoverable. A primary purpose of the present effort is to bring the telephone demand literature to the attention of a much wider audience, inside the telephone industry as well as outside. Indeed, the literature is so large that to attempt to cover everything at once would almost certainly invite reader discontent. The review will accordingly be approached in stages. Following an introduction and overview (which occupies this chapter), I shall proceed to a discussion of the theory of telephone demand in Chapter 2, then to a survey of the empirical literature in Chapters 3 and 4, and, finally, to an evaluation and critique in Chapter 5.

A READERS' GUIDE TO THE LITERATURE

The bibliography at the end of this monograph lists a substantial number of papers and monographs that contain at least passing reference to telephone demand.[1] Much of the list takes the form of unpublished papers and memoranda written by employees of the telephone companies, and although most of these studies are not viewed as proprietary, very few are known outside of their immediate environs. Consequently, while the literature of telephone demand is in fact quite large and also rich in content, the actual published literature is small. As one should imagine, much of the research on telephone demand is centered in the Bell System—specifically, at AT&T General Departments, AT&T Long Lines, the Bell Laboratories, and the Bell System operating telephone companies. After the Bell System, Bell Canada and the Trans Canada Telephone System have supported the largest research efforts. Academic economists have made a number of contributions, but in comparison with the literature on electricity demand, for example, their output has been rather modest.

In the studies emanating from the telephone industry, the focus is largely as one would expect. AT&T Long Lines has developed models for interstate toll demand and international demand, while the Bell System operating telephone companies have concentrated on intrastate toll demand. The Trans Canada Telephone System has also focused on toll demand. Interestingly enough, Bell Canada has undertaken the widest range of studies on a sustained basis. But this is not surprising since Bell Canada functions both as a local operating company and as a carrier of interprovince and international calls.

In terms of the quantity of research, toll demand has received by far the most attention, and local-service demand has received relatively little by comparison. While this may seem a bit surprising in view of the substantial revenues generated by local service, the reason is not hard to find. The traditional view in the telephone industry has been that the price elasticity for local service is very inelastic (i.e., very close to zero), so that an elaborate economic model of local service is not needed in order to forecast the impact of rate changes. Once it is decided how much additional revenue is to be obtained from local service, the rates are set accordingly. With toll demand, however, a nonzero price elasticity is known to

---

[1] Throughout this monograph, telephone and telecommunications demand will be used interchangeably, as will network and system.

exist, and an econometric model of toll demand has been recognized as an appropriate instrument to use in estimating its impact.

Since intrastate rates are regulated by state Public Utility Commissions, intrastate toll demand has accordingly received a great share of attention. Models for intrastate toll demand now exist for nearly all states.[2] AT&T Long Lines has the Long-Distance–Interstate (LDI) model which has been used before the FCC,[3] and also has some thirty-odd models of international demand.[4] And, in Canada, Bell Canada and the Trans Canada Telephone System have developed a stable of toll models corresponding to the various toll markets that these companies serve. Finally, other studies that examine toll demand include Feldman (1976, U.S. intrastate), Davis et al. (1973, U.S. intrastate and interstate), Deschamps (1974, Belgium), Dobell et al. (1972, Canada), Rea and Lage (1978, international), Lago (1970, international), Larsen and McCleary (1970, U.S. interstate), Waverman (1974, Canada, Great Britain, and Sweden), Wert (1976, U.S. interstate), and Yatrakis (1972, international).

In the next chapter we shall emphasize the distinction between the demand for access to the telephone system and the demand for use of the system once access has been acquired. Access demand corresponds to the demand for telephone lines, while use demand corresponds to the demand for calls. In this framework, there is no distinction, in principle, between toll calls and local calls, since both represent use as opposed to access. Consequently, basic exchange service and local use should in principle be modeled separately. In most localities, however, local-call service is typically tied to access, which blurs the distinction between the demand for access and the demand for local use.

The problems that this creates in analyzing local-service demand have been handled in a number of ways. In terms of the framework to be developed in the next chapter, the ideal procedure would be to specify separate equations for the number of telephones and local use. However, the only instances of this in the literature are the studies of Pousette (1976) and Waverman (1974), both of which use data for Sweden. In the United States and Canada an equation

---

[2] The reference here is to intrastate toll models developed by Bell System operating telephone companies.

[3] See Betteridge (1973). The LDI model has also been used in a number of filings before state Public Utility Commissions. Work is currently under way to replace the LDI model and a brief description of this work is in the *Postscript* in Chapter 6.

[4] See Craver (1975, 1976a, 1976b, 1976c, 1976d, 1977).

for local use cannot be specified for most localities because the price of local use is combined with the price of access. There are only six U.S. studies that focus explicitly on the demand for local calls, namely, Beauvais (1977), New York Telephone Co. (1976a, 1976b), and Pavarini (1975, 1976b, 1978). Beauvais analyzes the demand for local calls in three exchanges in central Illinois, while the NYT studies analyze the demand for additional message units in New York City. Pavarini, in contrast, is concerned with predicting the impact on local calling of a conversion from flat to measured-rate service.

The remaining North American studies of local-service demand estimate models for price-deflated local revenues. However, as will be suggested in Chapter 3, the dependent variable in these models is to be identified more with the demand for access than with the demand for local use. Hence these studies, which include Davis et al. (1973), Dobell et al. (1972), New York Telephone Co. (1976a, 1976b), and Southern New England Telephone Co. (1977a), are probably more appropriately grouped with the studies that focus on the demand for access.

The demand for access has received more attention than the demand for local use (but not nearly as much as toll demand). We have already mentioned Pousette (1976) and Waverman (1974), both of whom focus explicitly on the access/use distinction using data for Sweden. Waverman also estimates models for Canada and Great Britain. The measure of access for Waverman is the number of main-station telephones, which is also the case for Pousette, except that the quantities that Pousette actually explains are new connections and disconnections. Other studies that contain equations explaining the number of telephones include Alleman (1977), Feldman (1976), Perl (1978), and Rash (1971). Perl's study, which uses data from the 1970 Census, differs from the others in that it employs a broader definition of access. Access is usually defined as the number of main-station telephones for residential customers and the number of main stations plus PBX extensions for business customers, but Perl defines access in terms of having a telephone on the premises or else having access to one nearby.

Besides basic-exchange and long-distance services, telephone companies provide a variety of other services, but these have received only limited attention in the literature. The demand for customer terminal equipment other than the main station—the so-called vertical services—has been analyzed by Feldman (1976) and also by Huang (1975), Kearns (1978), New York Telephone Co. (1976b), Reitman (1977), and Southern New England Telephone

Co. (1977a). Two studies, Black and Tryon (1976) and Southern New England Telephone Co. (1977a), have analyzed the impact on new connections of service-connection and installation charges.[5] The demand for wide area telephone service (or WATS) and private lines has been addressed by Feldman (1976), and Davis et al. (1973), and Subissati (1973), who analyzes data for Bell Canada. As already noted, international telephone demand has been analyzed at AT&T Long Lines, Bell Canada, and the Trans Canada Telephone System, and also by Rea and Lage (1978), Lago (1970), Naleszkiewicz (1970), and Yatrakis (1972). Coin stations have been analyzed in the Feldman study and also by Brubacher and Hiscock (1978). Finally, yellow-pages advertising has been addressed by Wang (1976).

Most of the literature on telephone demand is of very recent origin. The first major quantitative model of telephone demand was the LDI model of interstate toll demand developed by AT&T Long Lines in 1967, but econometric modeling of telephone demand did not catch on until the early 1970s. Davis et al. (1973) and Dobell et al. (1972) represent two of the first efforts, but in each case demand is only part of a larger focus. An econometric model of AT&T which includes modules for demand, production, and financial and capital markets is developed by Davis et al., while production and investment, as well as demand, are analyzed by Dobell et al.[6] As already mentioned, toll demand (especially intrastate toll demand) bulks largest in the literature. However, the analysis of intrastate toll demand is a very recent phenomenon and reflects the introduction of econometric models into rate hearings before a substantial number of state Public Utility Commissions.

The studies in the literature vary considerably in terms of comprehensiveness. Most of the existing studies focus on just one component of demand, and only a few studies are truly comprehensive. The most comprehensive study is the one by Feldman (1976), which estimates equations for access, use, and type of call for business and residential customers, WATS, and coin stations. Other comprehensive studies include Davis et al. (1973), Dobell et al. (1972), New York Telephone Co. (1976a, b), Pousette (1976),

---

[5] These changes are usually referred to as non-recurring charges.

[6] The Dobell et al. study was part of a larger study of the communications sector in Canada [Dobell et al. (1970)] that was commissioned by the Department of Communications. The telegraph and broadcasting industries were examined, as well as the telephone industry. The work was done in the spring and summer of 1970 at the Institute for Policy Analysis at the University of Toronto. I mention this bit of history because the analysis of telephone demand was done under my direction.

Southern New England Telephone Co. (1977a), and Waverman (1974). These studies model most of the services that are provided by telephone companies—that is, access, local service, and toll. However, one cannot discuss the size of modeling efforts without mentioning the pioneering LDI model of AT&T Long Lines. Although the LDI model is restricted to interstate toll demand, it is far and away the most extensive telephone demand model in existence.[7] The LDI model allows for three customer classes, fifteen lengths of haul, two types of calls, three types of day, and twenty-five times of day, or 6,750 separate sectors.

Thus far in this "review of the troops," the focus (at least implicitly) has been on studies that are primarily concerned with quantifying price and income elasticities of demand. However, there are a number of studies whose purpose is primarily descriptive, as opposed to analytical, and it is from these studies that many of the "stylized facts" of telephone demand emerge. These studies can be conveniently grouped under three headings: (1) the studies that focus on the characteristics of toll traffic [Gale (1971, 1973), Harbus (1975)], (2) the studies that focus on access to the telephone system and the characteristics of local traffic [Garfinkel (1977a, 1977b, and forthcoming), Pavarini (1976a)], and (3) the studies that focus on the socioeconomic characteristics of residential customers [Brandon (forthcoming), Collins and Infosino (1978), and Southern New England Telephone Co. (1977c)]. The most important findings from these studies will be summarized in Section 2 below.

Three other descriptive studies warrant brief mention, an unpublished AT&T study and two Norwegian studies, Bø, Gaustad, and Kosberg (1976) and Myskja and Walmann (1973). The unpublished AT&T study examines the impact on business usage in three New Jersey Bell wire centers arising from the mandatory conversion in October 1975 to measured service from flat-rate service. Included among the findings are that business usage fell and that the reaction of single-line customers was stronger than that of multi-line customers. The Norwegian studies provide information on dialing time, abandoned calls, waiting time when there is no answer or there is a busy signal, and the time required for the called party to answer.

---

[7] For a description of the LDI model, see Betteridge (1973). See also Lowry (1976). The LDI model, it should be noted, is only partially econometric, since most of its parameters are calibrated through simulation techniques.

A striking feature of the telephone demand literature is the gulf between theory and empirical application. For all practical purposes, there are two literatures, a small theoretical literature that dates mostly from 1973, and the much larger empirical literature that we have been discussing. As we shall see in the next chapter, most of the theoretical literature on telephone demand consists of a series of papers published in 1973–1975 in the *Bell Journal of Economics*, notably Artle and Averous (1973), Squire (1973), Rohlfs (1974), Von Rabenau and Stahl (1974), and Littlechild (1975).[8] So far these studies have had only limited impact on applied econometric work, although this is not to say that the empirical literature is void of theory. Most of the empirical work has been guided by the general precepts of the theory of consumer demand. Demand functions are specified in which income and price appear as arguments, and in the models of international demand, the LDI model, and many of the models of Feldman (1976), prices of substitutes and complements are also included. There is also a great deal of attention given to habit formation, and a number of models include the stock of telephones as a predictor. Missing from the existing empirical models, however, is attention to the rather singular aspects of telephone demand.[9]

Still another feature of the empirical literature is the high level of aggregation, for the unit of observation in most studies is a state, province, or country as a whole. Studies with the observational unit below the state or province number only a few. Alleman (1977) analyzes data for cities, while Beauvais (1977), Perl (1978), and Pavarini (1975, 1976a, 1976b, 1978) all analyze data that refer to individual households.

THE STRUCTURE OF TELEPHONE DEMAND

The telephone serves a variety of wants and needs, and its demand has a number of distinguishing characteristics. Some of these characteristics are intrinsic to the basic nature of the telephone and the

---

[8] Alleman (1977) develops a model for access demand that is based on Squire (1973). This is true also of Mitchell (1978b), although Mitchell does not undertake any new econometric analysis.

[9] The nearest thing to a standard model in the empirical literature is the one that is used by a majority of Bell System operating telephone companies in analyzing intrastate toll demand. The number of toll messages (or price deflated toll revenues) is related to disposable income, the price of a toll call, the stock of telephones (as a measure of the size of the "market"), and the lagged dependent variable (as a measure of "habit").

system of which it is a part, but there are other characteristics that arise empirically and which at present do not lend themselves to theoretical explanation. The consumption externalities associated with the size of the telephone system illustrate the former, while the empirical finding that the mean duration of a call varies directly with the length of haul illustrates the latter.

The empirical characteristics of telephone demand can be grouped under three headings: (1) sociodemographic characteristics, (2) time, distance, and customer characteristics, and (3) characteristics relating to price and income elasticities of demand. The characteristics to be listed will be viewed as descriptive features of telephone demand, and most of the ones to be noted under the first two headings derive from descriptive studies. However, the characteristics under the third heading are all drawn from studies whose primary goal is to estimate behavioral models of demand.

Sociodemographic Characteristics

Four recent studies in the Bell System have focused on sociodemographic characteristics of residential customers. The studies have analyzed data for Cincinnati Bell, Illinois Bell, Pacific Telephone and Telegraph Co., and Southern New England Telephone Co.[10] The Cincinnati Bell study was conducted between March 1975 and January 1976 on a sample of 731 households served by eight wire centers in Cincinnati. Data were obtained on the average number of local calls per line, and sociodemographic information was solicited in a mail questionnaire. Two hundred ninety-three households responded fully to the questionnaire. The Illinois Bell study relates both toll and nontoll calling characteristics for a sample of Chicago households to several sociodemographic factors. Data on telephone usage (number of calls, duration, time of day, and length of haul) were collected between October 1972 and April 1974 for 849 households attached to eleven wire centers. Sociodemographic data for the households were obtained in a mail/telephone/in-person questionnaire in March 1973. Six hundred twenty-three households responded to the questionnaire.

The data for the Pacific study were collected from a sample of 1,719 households served by ten wire centers in or near Los Angeles and San Francisco between May 1972 and September 1973. In-

---

[10] The Cincinnati Bell study is described in Collins and Infosino (1978), the Illinois Bell study in Brandon (forthcoming), the Pacific Bell study in Infosino (1976), and the Southern New England study in SNET (1977c).

formation was obtained on the number of calls per account per day, as well as on sociodemographic characteristics in a mail questionnaire. Seven hundred five households responded fully to the questionnaire.[11] Finally, the SNET study obtained data on daily minutes of use for 1,748 households served by nine wire centers in Connecticut between November 1974 and October 1975. Sociodemographic information was obtained in a mail questionnaire, to which 1,168 of the 1,748 households in the original sample responded.[12] One of the findings from these four studies is the suggestion that local-calling behavior of nonwhite households differs from that of white households.[13] Specifically, nonwhite households appear to make more local calls (all four studies), to speak longer (Illinois Bell), and to make fewer calls to the suburbs (Illinois Bell), which suggests a geographically tighter community of interest on the part of nonwhite households. The higher calling frequency of nonwhite households shows up at all levels of income in the Illinois Bell study, but only for households with less than $10,000 income in the SNET study. The Illinois Bell study also indicates that the peak calling frequency is in the evening for nonwhite households; however, for white households, there are twin peaks, one in mid-afternoon and the other at the same time in the evening as for nonwhite households.

Another finding is that local telephone usage is directly related to the size of the household (Cincinnati, Illinois Bell, and Pacific), and the same three studies also indicate that local usage is lower in households headed by a person aged sixty-five or more. Finally, the Cincinnati and Pacific studies taken together suggest that regional differences in local-calling behavior exist that cannot be accounted for by race, income, or household age or size, or by other variables used in individual studies, such as telephone density, sex of household heads, and so on. In general, however, the results reported in these studies must be viewed as suggestive, rather than definitive. A lot more research is required before there will emerge a complete picture of the impact of sociodemographic forces on calling patterns.

---

[11] Data for business customers were also collected in the studies, but only the Pacific data have been analyzed, and this on only a preliminary basis.

[12] The socio-demographic information obtained in the mail questionnaires was augmented by data from the 1970 Census in all of the studies except the one by Illinois Bell.

[13] Nonwhite in the Illinois Bell and Cincinnati Bell studies refer specifically to blacks. In the other two studies, nonwhite includes hispanics and orientals as well as blacks.

Time, Duration, Distance, and Customer Characteristics

Telephone companies view a telephone call as having five dimensions:[14]

1. The time-of-day and day-of-week that the call is made;
2. The type of call (DDD, person-to-person, etc.);
3. Duration of the call;
4. Length-of-haul of the call;
5. Customer class making the call.

Six studies, Garfinkel (1977a, 1977b, forthcoming), Pavarini (1976a), and Gale (1971, 1973), have examined the distribution of calls along these dimensions. Both Garfinkel and Pavarini focus on local use, while Gale focuses on interstate toll demand.[15]

The structural characteristics associated with local use reported in the Garfinkel and Pavarini studies include

1. Business customers make more local calls per account per day than residential customers; business customers also have more account-to-account variation in calling rates.
2. Holding times (i.e., call durations) are longer for residential accounts than for business accounts.
3. Flat-rate business classes have higher calling rates, in the same area, than measured-rate business classes.

---

[14] The price of a long-distance call varies according to the time-of-day and day-of-week, duration, and length-of-haul, but not by class of customer. In general, the price of local calls does not vary by time-of-day or day-of-week, but may vary by duration and length-of-haul, depending upon whether the price of local service is flat-rate or measured-rate.

[15] The data reported on in the Garfinkel studies were collected during the last seven months of 1972 in ten #1 ESS exchanges in California. Two thousand four hundred forty-one accounts were represented in the data analyzed, 1,292 residential accounts and 1,149 business accounts. Both flat-rate and measured-service accounts are represented in the sample. The data analyzed by Pavarini were collected in sampling studies conducted in seventy-nine Bell System wire centers in ten states between June 1972 and December 1975. A total of 7,253 flat-rate local-service accounts, both residence and business, are represented in the sample. The data analyzed in both the Garfinkel and Pavarini studies were mostly collected by the Subscriber Line Use Study (SLUS) System, which is a mechanized data collection system developed at Bell Laboratories and offered to the Bell System operating telephone companies by AT&T. Finally, the data analyzed by Gale were obtained from the Bell System MASP data files. For a description of the MASP data, see AT&T (1978).

4. A gamma distribution (or a power-transformed normal distribution in the case of Pavarini) provides a reasonable description of local-calling rates within each class of service.[16]
5. Mean local-calling rates for both business and residential customers appear to be distributed normally across wire centers.

Gale, in contrast with Garfinkel and Pavarini, focuses on the structural characteristics of toll use. The data analyzed by Gale were obtained from the Bell System MASP data files and were collected between 1966 and 1971. Most of Gale's emphasis is on interstate messages, but some intrastate messages are also analyzed. Some of the more interesting of Gale's findings regarding the distribution of toll messages by time-of-day [Gale (1973)] are as follows:

1. The percentage of calls for a typical weekday by residential customers that are made in the "daytime" as opposed to "nighttime," decreases with mileage.[17]
2. The day of the week makes a large difference in the distribution of calls by time-of-day for business traffic, but only a moderate difference for residence traffic. The typical business weekday traffic has a morning peak between 10:00 and 11:00 A.M. and an afternoon peak between 2:00 and 3:00 P.M. The morning peak is also the day's peak. Saturday has a morning peak virtually identical to the weekday morning peak, but no afternoon peak. Sunday has only a small morning peak, and an evening peak.
3. For residence traffic, there is always a morning peak, an afternoon dip, and the day's peak in the evening. The magnitude and location of the morning peak changes between weekdays and the weekend, and the magnitude of the evening peak decreases on the weekend, although the peak hour remains 7:00 to 8:00 P.M.
4. Collect calls were distributed much like paid calls, but billings to third parties show unusual patterns for both business and residence. Third-party billings for business customers show a secondary evening peak, which suggests that much of this traffic

[16]The gamma distribution is a two-parameter distribution that is unimodal and skewed to the right. A power-transformed normal distribution simply means that a random variable raised to some specified power has the normal distribution.

[17]Since the differential between daytime and nighttime rates increases with mileage, this result probably also reflects a price effect.

may be personal rather than commercial. On the other hand, third-party billings to residential accounts show the typical business pattern of morning and afternoon peaks, which suggests that much of this traffic may be commercial.

5. Finally, person-to-person calls show a more businesslike pattern than do station-to-station calls. For business traffic, the daytime peaks are more pronounced for person-to-person than for station-to-station calls, while for residence traffic person-to-person calls follow the business pattern of morning and afternoon peaks.

In his 1971 paper Gale examines the variation in the average length of calls with respect to such factors as mileage, class of customer, and time-of-day. The data analyzed refer to interstate toll calls for March 1969, and the findings include

1. The mean duration of a call is longer, other things being constant, the more distant is the call—or, "the longer the haul, the longer the call," as the Bell System saying goes.
2. The mean duration of a toll call from residential accounts is longer than from business accounts. The same is true, it will be recalled, for local calls.
3. Weekday calls are shorter, on the average, than weekend calls.
4. Calls between 7:00 P.M. and midnight are longer, on the average, than during the daytime.[18]
5. The mean duration of person-to-person calls is longer than station-to-station calls.
6. Collect calls and calls billed to third parties are longer, on the average, than paid calls.

The Structure of Price and Income Elasticities of Demand

Most of the focus in the empirical literature is on the estimation of price and income eleasticities of demand. These elasticities and the models that underlie them will be reviewed in Chapters 3–5, and, as we shall see, our knowledge in most areas is still rather rudimentary. However, a number of points are pretty well established. In particular:

1. Price and income elasticities of demand are definitely different from zero. While there still remains a great deal of uncertainty

---

[18] Since nighttime rates are lower than daytime rates, this difference will also contain a price effect.

as to the actual magnitude of the elasticities, any idea that telephone services are consumed without regard to the prices of telephone services or the level of income must be dismissed.

2. On the whole, income elasticities are quite a bit larger (in absolute value) than price elasticities. In general, it seems safe to conclude that telephone demand in the aggregate is elastic with respect to income and inelastic with respect to price.
3. The price elasticity of the demand for access to the telephone system (i.e., the demand for main-station telephones plus business PBX extensions) is very small, but nonetheless definitely different from zero.
4. In general, the empirical estimates of price elasticities establish that the price elasticity becomes larger (in absolute value) as one goes from access to local calls to short-haul toll calls to long-haul toll calls to international calls. The same pattern also appears to hold for income elasticities.
5. Finally, telephone demand is characterized in most instances by rather strong inertia, which is to say that adjustments in telephone demand to changes in price or income are not instantaneous but take place over a period of time. As a consequence, price and income elasticities of demand are smaller in the short run than in the long run.

Chapter 2

The Theory of Telephone Demand

Meaningful empirical analysis of telephone demand must be accompanied by theory. While it is true that most existing studies are guided by the general canons of neoclassical demand theory, it is also true that neoclassical theory does not have anything very specific to offer as regards telephone demand. The standard statement that demand depends upon price and income can only take one so far. As will be discussed in this chapter, the telephone possesses a number of highly individual features that need to be considered in modeling its demand. These features have been examined at the theoretical level in several places, but the analyses are highly specialized, and the contributions involved have largely been ignored empirically.

The purpose of this chapter is to develop a theoretical structure for analyzing telephone demand that recognizes the unique features of the telephone and the network of which it is a part. The theoretical structure to be developed will provide a framework for evaluating existing empirical research and will also assist in Chapter 5 in organizing an agenda for future research.

THE TELEPHONE AND THE SERVICES THAT IT PROVIDES

A completed telephone call requires the participation of a second party, and the utility of this party is accordingly affected. The second party is presumably also benefited, but whether or not this is the case is beside the point. The point is that a completed call necessarily impinges on a second party, and an externality is thereby created. A second type of externality arises when a new subscriber connects to the system. Connection of a new subscriber confers a benefit on existing subscribers because the number of telephones that can be reached is increased. In this case, the externality is an *access* (or *system*) externality, whereas, in the first case, the externality is a *call* (or *use*) externality.

The access externality gives the telephone the dimension of a public good, since the benefit that a new subscriber confers on existing subscribers is shared in common.[1] The public-good aspect of the telephone has been the subject of a great deal of recent attention, notably in a series of articles published in the *Bell Journal of Economics.*[2] However, in this review, our interest is not with public goods per se, but rather with the implications of the access and call externalities for the theory of telephone demand and the empirical models to be derived therefrom.

A further complication in modeling telephone demand arises from the fact that benefits come not only from completed calls, but also from calls that may not be made. When an individual subscribes to the telephone system, he in effect purchases options to make and receive calls. Some of these options will be exercised with certainty, while others will not be. This is because many calls will only be made contingent upon particular states of nature whose realization is essentially random and therefore not known at the time that access is purchased. Calls of an emergency nature, such as for fire, police, or ambulance, are obvious cases in point, but compelling urgency is not the only determinant. *Option demand* is thus an important characteristic of telephone demand.[3]

Before turning to the formal aspects of the theory of telephone demand, it will be useful to put the foregoing into perspective by summarizing several of the more important points that must

---

[1] The call externality also carries a public-good dimension, but the effect is highly localized in comparison with the access externality since only one other party is affected.

[2] See Artle and Averous (1973), Squire (1973), Von Rabenau and Stahl (1974), Rohlfs (1974), and Littlechild (1975). Artle and Averous, Von Rabenau and Stahl, and Rohlfs mainly focus on two questions: (1) the optimal number of telephones in an economy, and (2) self-sustaining endogenous growth induced by the access externality. Squire and Littlechild, on the other hand, are primarily concerned with optimal pricing. In each paper, however, the public-good nature of the telephone is of major importance.

[3] The concept of option demand is well known in the literature on conservation and exhaustible natural resources, but, except for a brief passing reference in Squire (1973), the concept has not found its way into the literature on telephone demand.

be kept in mind in building a meaningful theory of telephone demand:

1. The point of departure in modeling telephone demand ought to be the distinction between demand for *access* to the system and demand for *use* of the system once access is obtained.
2. The demand for access is complicated by both the access and call externalities, while the demand for use is complicated by the call externality. The presence of these externalities means that, contrary to what is ordinarily assumed, *preferences are interdependent across subscribers.*
3. Also, as just discussed, option demand is an important component of telephone demand, for a subscriber is willing to pay something for the option to make and receive calls even though the calls may never be made or received.
4. Finally, a telephone call should be distinguished by type (station, person, collect, etc.), time-of-day, distance, and duration.

RECENT CONTRIBUTIONS TO THE THEORY OF TELEPHONE DEMAND

The problems associated with modeling telephone demand have been clarified substantially in the last few years in a series of papers in the *Bell Journal of Economics* by Artle and Averous (1973), Von Rabenau and Stahl (1973), Squire (1973), Rohlfs (1974), and Littlechild (1975). The contributions of these authors, beginning with Artle and Averous, will be presented and discussed in this section.

Artle and Averous consider a population consisting of N individuals that are a group in the sense that they all enjoy talking to one another. It is assumed that, over a period of time, each individual in the population converses once with every other individual either by telephone or face to face. In addition, it is assumed that if an individual is not connected to the telephone system as a subscriber, he has no access to the system at all (i.e., there are no pay phones and there is no use of another's phone). This, accordingly, defines two mutually exclusive and exhaustive subsets of individuals, say G_0 and G_1, where G_0 denotes the subset without access to the system.

The utility function for the ith individual is assumed to be given by

$$U^i = U^i(x^i, q^i), \qquad (1)$$

$$q^i = \begin{cases} q \text{ for all } i \text{ in } G_1 \\ 0 \text{ for all } i \text{ in } G_0. \end{cases} \tag{2}$$

In (1), q^i stands for access by the ith individual to the telephone system, denoted by q,[4] while x^i denotes the ith individual's consumption of other goods. For simplicity, the goods included in x are all assumed to be private.

Since the property of access or no access is dichotomous, it can also be represented by a dummy variable δ^i, where $\delta^i = 1$ if the ith individual has access to the telephone system and $\delta^i = 0$ if the individual does not have access. Consequently, an equivalent formulation of the utility function in (1) is

$$U^i = U^i\,(x^i, \delta^i q^i), \tag{3}$$

where $q^i = q$ for all i and

$$\delta^i = \begin{cases} 1 \; i \epsilon G_1 \\ 0 \; i \epsilon G_0. \end{cases} \tag{4}$$

Artle and Averous next assume that all individuals connected by telephone (i.e., the members of G_1) consume the same bundle of private goods, x^1. Similarly, it is assumed that the individuals not connected to the telephone system (i.e., the members of G_0) are also alike in that they, too, consume a common bundle of private goods, x^0. With these assumptions the utility functions for individuals with and without a telephone can be written as[5]

$$U^1 = U^1\,(x^1, \delta q), \qquad \delta = 1 \tag{5}$$

$$U^0 = U^0\,(x^0, 0), \qquad \delta = 0 \tag{6}$$

At this point, Artle and Averous postulate the presence of an omniscient planner who is equipped with a social-welfare function

---

[4] The units in which q is measured will be defined below.

[5] Artle and Averous actually assume that U^1 and U^0 have the same functional form.

that embraces the well-being of all N individuals in the population. This omniscient planner is also provided with a social-production-possibility function that reflects the technological constraints imposed on the behavior of the N individuals. Finally, it is assumed that q (whose units have not previously been defined) is measured by the number of telephones, which implies, then, that q of the N individuals in the population are connected by telephone, while $N - q$ individuals are outside of the telephone system.[6] Question: What is the optimal value of q?

Let W denote the social-welfare function and F the social-production-possibility function. Artle and Averous assume that W has the form

$$W = W[qU^1, (N-q)U^0]. \tag{7}$$

The question posed can be answered by maximizing the function

$$V(x^1, x^0, q) = W[qU^1, (N-q)U^0] - \lambda F(x,q) \tag{8}$$

with respect to x^1, x^0, and q, where λ is the LaGrangian multiplier associated with the social-production-possibility function,

$$F(x,q) = 0, \tag{9}$$

and where

$$x = qx^1 + (N-q)x^0. \tag{10}$$

Assuming that an interior solution exists ($1 < q < N$), the necessary conditions for an optimum will be given by[7]

$$\frac{\partial V}{\partial x^1} = W_1 q U_x^1 - \lambda F_x q = 0 \tag{11}$$

$$\frac{\partial V}{\partial x^0} = W_0 (N-q) U_x^0 - \lambda F_x (N-q) = 0 \tag{12}$$

$$\frac{\partial V}{\partial q} = W_1 (U^1 + q U_q^1) - W_0 U^0 - \lambda F_q = 0 \tag{13}$$

$$\frac{\partial V}{\partial \lambda} = F(x,q) = 0. \tag{14}$$

---

[6] In the analysis that follows, q will be treated as though it varies continuously.

[7] In equations (11) - (13), W_1 denotes $\partial W/\partial\ (qU^1)$, W_0 denotes $\partial W/\partial$ $(N-q)U^0$, F_q denotes $\partial F/\partial q$, and U_x^i denotes $\partial U^i/\partial x$.

From (11) and (12), we obtain the following optimality condition:

$$W_1 U_x^1 = W_0 U_x^0, \tag{15}$$

which simply states that the last unit of private consumer goods yields the same welfare, as measured by the social-welfare index, independent of the individual to which it is allocated. From (11), (12), and (13) we can derive

$$\left(\frac{U^1}{U_x^1} - \frac{U^0}{U_x^0}\right) + q\,\frac{U_q^1}{U_x^1} = (x^1 - x^0) + \frac{F_q}{F_x}. \tag{16}$$

The interpretation of the condition in (16) is as follows. To begin with, the second term on the right-hand side is the usual marginal rate of transformation in production, which, in this case, represents the marginal amount of x that must be given up in order to produce (and service) one more telephone. The second term on the left-hand side gives the marginal rate of substitution in consumption between telephone and other goods. It is seen that the marginal rate of substitution is multiplied by q, the number of individuals that have access to the telephone system. *This result shows that access to the telephone system is indeed a public good.*

As regards the two terms in parentheses, the term on the right-hand side measures the change in other goods that occurs when an individual enters the telephone system on the margin. Prior to entering the system, x^0 of other goods was consumed; after entry, x^1 of other goods is consumed. The difference between the two, if negative, reduces the marginal rate of transformation of telephones for other goods. If the difference is positive, this difference must be produced, which will increase the marginal rate of transformation. In the same vein, the term in parentheses on the left-hand side shows the net difference in total welfare by having the marginal individual switch from the market basket $(x^0, 0)$ to the basket (x^1, q). Since the terms in expression (16) are defined in units of x, both U^1 and U^0 are measured in the respective marginal utilities of x.

The Artle-Averous analysis makes two important contributions to the development of a theory of telephone demand: (1) the public-good aspect of the telephone system is demonstrated with clarity, and (2) the dichotomous access/no access decision is formulated in a way that allows the use of conventional calculus methods in deriving the optimal size of the telephone system. However, as a basis for a theory of telephone demand, the Artle-Averous analysis is incomplete. Prices and income are subsumed in the social-production-possibility function, and telephone use, as opposed to access, is not really analyzed. The demand for use is in effect assumed away by the Artle-Averous assumption that each individual in the population communicates once with every other individual over some period of time, either by telephone or face-to-face contact. Still a further limitation with the Artle-Averous analysis is that the focus is on the optimal size of the system, rather than on individual demand.

Rohlfs (1974) extends the Artle-Averous analysis in several important respects, and in doing so overcomes some of the problems just noted. Like Artle and Averous, Rohlfs, too, is concerned with the determination of an equilibrium number of telephones in a population, but he gives much greater attention to the demand side of the market. The public-good aspect of the telephone continues to be central to the analysis, but the focus is on individual demand, and the omniscient planner is not employed. The market demand function for telephones is obtained by aggregating over individual demand functions, which are derived from utility maximization.

The key construct in Rohlfs' analysis is the concept of an *equilibrium user set*, which is defined as a set of telephone users that is consistent with all individuals (users and non-users alike) maximizing their respective utility functions. A basic finding of Rohlfs is that, for any given price, there are typically multiple equilibria:

> For example, a very small equilibrium user set may be consistent with utility maximization, since the smallness of the user set in itself makes the service relatively unattractive to potential users. However, a much larger user set may also be possible for the same population at the same price. In this case the largeness of the user set would make the services attractive and allow a high level of demand to be sustained. [Rohlfs (1974, p. 18)].

Rohlfs begins his analysis by defining (a la Artle and Averous) a set of binary variables:

$$q_i = \begin{cases} 1 \text{ if individual } i \text{ has a telephone} \\ 0 \text{ if not} \end{cases} \tag{17}$$

for $i = 1, \ldots, N$. It is assumed that there are s other goods in the economy (all of which are private). Next, it is postulated that each individual possesses a *pair* of utility functions:

$$U_i^0 = U_i^0\ (x_{i1}, \ldots, x_{is}) \tag{18}$$

$$U_i^1 = U_i^1\ (q_1, \ldots, q_{i-1}, q_{i+1}, \ldots, q_N, x_{i1}, \ldots, x_{is}), \tag{19}$$

where

U_i^0 = utility of individual i if the individual does not subscribe to the telephone system
U_i^1 = utility of individual i if the individual does subscribe to the telephone system
x_{ij} = consumption of (nontelephone) good j by individual i.

The utility function in (19), it will be noted, assumes that the utility of a telephone subscriber depends upon the other individuals who belong to the telephone system. This interdependence of preferences reflects the consumption externalities of the telephone discussed in Section 1.

Per usual, Rohlfs assumes that, for $k = 1, 2$,

$$\frac{\partial U_i^k}{\partial x_{ij}} \geqslant 0 \quad \text{for all } j \tag{20}$$

and

$$\frac{\partial U_i^k}{\partial x_{ij}} > 0 \quad \text{for at least one } j. \tag{21}$$

Moreover, it is assumed that

$$U_i^0 \leqslant U_i^1 \tag{22}$$

for all i and for all values of the x_{ij}'s. Finally, it is assumed that

$$\frac{\Delta U_i^1}{\Delta q_w} \geqslant 0 \tag{23}$$

for all $i \neq w$ and for all $q_1, \ldots, q_{i-1}, q_{i+1}, x_{ij}, \ldots, x_{is}$. This last assumption states that the addition of another subscriber to the telephone system will never cause an existing subscriber to disconnect.

In deriving the demand function for access, Rohlfs postulates a two-step maximization procedure. First, the maxima of U_i^0 and U_i^1 are evaluated with respect to $q_1, \ldots, q_{i-1}, q_{i+1}, \ldots, q_N$, $x_{i1}, \ldots, x_{is}$ subject to individual i's budget constraint. Denote these maxima by $\hat{U}_i^0$ and $\hat{U}_i^1$. The demand variable for individual i, q_i^d, is therefore defined as

$$q_i^d = \begin{cases} 1 \text{ if } \hat{U}_i^1 \geqslant \hat{U}_i^0 \\ 0 \text{ if } \hat{U}_i^1 < \hat{U}_i^0 \end{cases} \tag{24}$$

In order to concentrate on the interdependency in the telephone market, Rohlfs assumes income and the prices of all other goods to be constant, which allows i's *demand function* for access to the telephone system to be expressed as

$$q_i^d = q_i^d(r, q_1, \ldots, q_{i-1}, q_{i+1}, \ldots, q_N), \tag{25}$$

where r denotes the price of access. From earlier assumptions, it follows that q_i^d is nonincreasing in r—that is, an increase in r will never change q_i^d from 0 to 1, while a decrease in r will never change q_i^d. Finally, it follows from the assumption embodied in expression (23) that all q_i^d are nondecreasing with respect to all q_w ($w \neq i$).

Rohlfs defines an *equilibrium user set* as a set of telephone subscribers for which

$$q_i = q_i^d(r, q_1, \ldots, q_{i-1}, q_{i+1}, \ldots, q_N) \tag{26}$$

for all i. Thus, in equilibrium, all users demand access to the system, while nonusers do not. For fixed r, the equations in (26) comprise a system of N equations in N binary variables. However, Rohlfs shows that such a system does not in general have a unique solution. Indeed, in none of the simple models that Rohlfs investigates is the solution unique, except in the trivial case where the price of access is so high that there is no demand at all. The reason for this non-uniqueness is clearly the consumption externality of the telephone. For, as noted earlier, an equilibirum user set can be very small because the smallness of the system makes the service unattractive to potential subscribers. On the other hand, a much larger equilibrium user set is also possible since the largeness of the system makes the service more attractive to potential subscribers.[8]

An important implication of the existence of multiple equilibrium user sets is the possibility that a telephone system can grow over a period of time even though external factors (income, price, and the size of the population) do not change. If disturbed from an equilibrium with a small user set, the system might not return to that equilibrium, but could move to a new equilibrium with a much larger user set. Whether or not this would occur depends, as Rohlfs shows, on the direction of the disturbance and the nature of the disequilibrium adjustment process. However, the really important point is the role played by the consumption externality. For, as also stressed by Artle and Averous, endogenous growth can arise because of it.

Rohlfs' analysis makes the following contributions to the development of a meaningful framework for analyzing telephone demand:

1. By introducing the notion of an equilibrium user set, Rohlfs avoids the necessity of resorting to a social-welfare function in order to determine the equilibrium number of telephones in a population.
2. A major contribution is made in approaching the derivation of the demand function for access to the telephone network in two steps: utility is evaluated with and without access to the telephone system, and the demand for access is then determined by comparing the resulting levels of utility. Not only does this procedure seem descriptive of actual behavior, but it also provides a way of integrating the best existing work on access demand, which is based on a consumer's surplus analysis, into a utility framework.
3. Whereas Artle and Averous subsume the cost of a telephone into the social-production-possibility function, telephone price is an explicit argument in Rohlfs' analysis.

Despite these contributions, Rohlfs' analysis, like that of Artle and Averous, has some shortcomings. In particular, Rohlfs does not

---

[8] It should be noted that equation (26) defines equilibrium only with respect to the demand side of the market. It describes subscriber sets that are consistent with utility maximization at a given price of access, but these constitute only necessary, not sufficient conditions for an overall market equilibrium. The latter requires, in addition, that the subscriber set and the price of access be consistent with some specified model of supply behavior. [See Rohlfs (1974, p. 21).]

distinguish between access and use. Rohlfs employs just one price, which refers to access, as opposed to use. In this case, the customer would pay a flat charge per time period and make as many calls as desired. However, this ignores toll calls and those local calls for which there is an additional charge. Consequently, a meaningful theoretical framework for analyzing telephone demand must include a price for use (when use is priced), as well as a price for access.

To this end, we shall now turn our attention to an analysis first introduced by Squire (1973) that relates the demand for access to the demand for use in a consumer's surplus framework. Let q be a dichotomous variable that takes the value 1 if the consumer desires to subscribe to the telephone system and which takes the value 0 if the consumer desires not to subscribe. Let π denote the price of a call, r the subscription price of access to the telephone system, and μ the level of income of the consumer. Next, let Z denote the number of calls that the consumer would make if he were connected to the telephone system and let

$$\pi = g(Z,\mu) \tag{27}$$

denote the inverse demand function for calls.[9] Finally, let B_Z denote the amount that the consumer would be willing to pay in order to make the Z calls.

Under Marshallian conditions, B_Z will be given by [10]

$$B_Z = \int_0^Z g(z,\mu)dz. \tag{28}$$

The consumer's surplus (S_Z) associated with the Z calls will then be equal to

$$S_Z = B_Z - Z\pi. \tag{29}$$

Finally, it follows that the binary variable q will be determined according to

$$q = \begin{cases} 1 \text{ if } S_Z \geqslant r \\ 0 \text{ if } S_Z < r. \end{cases} \tag{30}$$

---

[9] For now, the prices of other goods are absorbed into the shape of g. Note, too, that the customer index i is suppressed.

[10] For convenience, we shall treat the number of calls Z as though this were a continuous variable.

In words, expression (30) states that a consumer will demand access to the telephone system if the benefits from belonging to the system (measured by the consumer's surplus associated with the calls that would be made) are at least as great as the cost of subscribing to the system (measured by r).[11] From expressions (27) to (30), it follows, then, that (ignoring externalities) the demand for access is a function of income, the price of calls, and the price of access that is,

$$q = q(r, \pi, \mu). \tag{31}$$

Thus, we see that the demand for access to the telephone system depends in a fundamental way on the demand for use of the system. In making this connection precise and rigorous—and especially in showing that the relationship flows from use to access, rather than from access to use—Squire makes a fundamental contribution to the theoretical analysis of telephone demand.

In arriving at expressions (30 and (31), access and call externalities (except as noted in footnote 11) have been ignored. We shall now indicate how these externalities can be integrated into the analysis. The discussion that follows is once again based upon Squire.

To begin with, let us distinguish between a *conceptual* market demand curve for calls and an *observed* demand curve. The conceptual demand curve refers to the relationhship between the total number of calls that a group of subscribers would make and the price of a call on the assumption that the number of subscribers is constant. However, because of consumption externalities, there is an entire family of conceptual demand curves. Moreover, only a single point on each member of this family can be observed. For if we were to increase the price of a call to all subscribers, there would be a reduction in both the number of calls and the number of subscribers.

The number of subscribers would be reduced because the increase in the price of a call will decrease the net benefits of being connected to the telephone system, and this will cause subscribers that were previously on the margin to disconnect. Moreover, because of the

[11] It should be noted that a consumer may desire access to the system even though no calls are made. In this case, $q(0,\mu)$ would measure the benefits of receiving calls. If $q(0,\mu) \geqslant r$, the consumer will demand access just to receive incoming calls. The benefits generated by incoming calls are clearly related to the call externality referred to earlier, and will be discussed below.

externality associated with the number of subscribers, the decreased system size will lead to a still further attrition of subscribers. The reduction in the number of subscribers means, therefore, that, for any given size of the system, the conceptual demand curve will be less elastic than the observed demand curve.

The foregoing is illustrated in Figure 2-1. The total number of calls in the system is measured along the horizontal axis, while the price of a call is measured on the vertical axis. Three members of the family of conceptual demand curves are drawn. These are labeled $\overline{N}_1$, $\overline{N}_2$, and $\overline{N}_3$, and correspond to a system consisting of N_1, N_2, and N_3 subscribers, respectively, where $N_1 < N_2 < N_3$.[12] That the curve corresponding to N_3 lies to the right of the curve for N_2, which is to the right of the curve for N_1, reflects the positive externalities generated by the telephone system.

The conceptual demand curves, as mentioned, cannot be observed. What can be observed is the one point on each curve that corres-

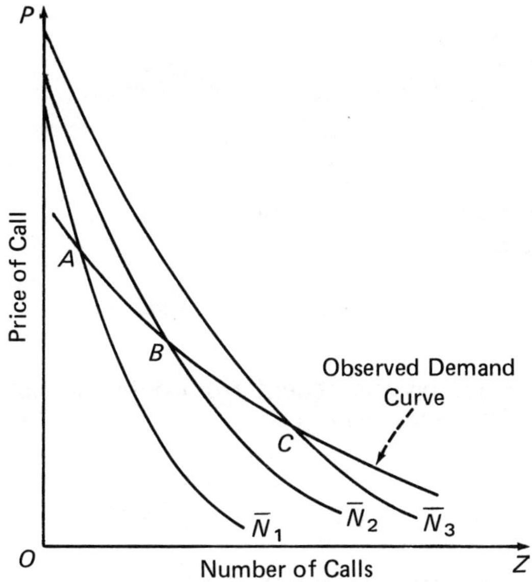

Source: Squire (1973), p. 517.

Figure 2-1. Relationship between the Conceptual and Observed Market Demand Curve for Calls.
Source: Squire (1973), p. 517.

[12]The bars over the *N*s are to remind us that the conceptual demand curve is derived with the number of subscribers held constant.

ponds to the equilibrium number of calls for that number of subscribers. (These points correspond to the equilibrium user sets discussed by Rohlfs.) The locus of these points yields the observed demand curve as drawn in the figure. As noted earlier, the elasticity of the observed demand function is greater than the elasticity of the conceptual demand function.

With a slight modification to our earlier notation, the benefit derived by the nth subscriber from outgoing calls can be written as

$$B_n(\overline{N}) = \int_0^{Z_n} g_n(z,\overline{N})dz, \tag{32}$$

where Z_n is the number of calls made by the nth subscriber and g_n is this subscriber's conceptual inverse demand function.[13] The total benefit from outgoing calls is obtained by summing over the N subscribers,

$$\sum_{n=1}^{N} B_n(\overline{N}) = \sum_{n=1}^{N} \int_0^{Z_n} g_n(z,\overline{N})dz. \tag{33}$$

If we define $G(Z, \overline{N}) = \sum_{n=1}^{N} g_n(Z_n, \overline{N})$, where $Z = \max(Z_n)$, as the market conceptual inverse demand function,[14] expression (33) can be written as

$$\sum_{n=1}^{N} B_n(\overline{N}) = \int_0^{Z} G(z,\overline{N})dz. \tag{34}$$

Finally, Squire defines the benefit from outgoing calls for the *marginal* (i.e., the Nth) subscriber as

$$B_N(\overline{N}) = \sum_{n=1}^{N} B_n(\overline{N}) - \sum_{n=1}^{N-1} B_n(\overline{N}). \tag{35}$$

The foregoing describes the benefits associated with outgoing calls, but what about incoming calls? Squire states (p. 518): "When one's telephone rings, one has no idea of the possible benefit of the incoming call." Presumably, the expected benefit is positive, but there is no guarantee that this will be the case, for the call

---

[13] Income has been absorbed into the shape of g_n.

[14] The three conditional demand curves in Figure 2-1 correspond to $G(Z, \overline{N})$ for $\overline{N} = \overline{N}_1, \overline{N}_2, \overline{N}_3$.

may be a nuisance call or the caller may be someone with whom the callee would prefer not to speak. However, as Squire notes, the benefit of an incoming call can, in principle, be measured by asking how much a person would be willing to pay in order to be allowed to answer the call. This reduces the problem to asking for the certainty-equivalent value of an uncertain return.

Specifically, let us assume, following Squire, that the possible benefit (i.e., return) from an incoming call can be represented by a probability distribution, where the expected return is measured by the mean of the distribution (μ) and the risk associated with the return by its variance (σ^2). Assume that a subscriber is willing to accept a greater risk (i.e., is willing to accept a greater uncertainty of the possible benefit of an incoming call) only in return for a larger expected benefit. We can than define a curve, as in Figure 2-2, that shows the nth subscriber's tradeoff between risk (i.e., higher variance) and expected benefit.[15] The point b_n on the μ-axis measures the certainty equivalent of the uncertain benefit of the incoming call. Hence, in this case, the subscriber is willing to pay up to b_n to answer the phone. We can therefore interpret b_n as the benefit from an incoming call.[16]

Squire observes:

> We can deduce something about the probable value of b_n from observed behavior. In the first place, most subscribers do answer their phones, which operation presumably involves some private cost in terms of effort expended. (This private cost must be quite high in the case of the subscriber who gets out of his shower to answer his phone!) In some cases, the private cost may be negligible, but in general we can assume that it is not negative (i.e., subscribers do not derive utility from lifting the receiver), from which observation we can deduce that $b_n \geqslant 0$. This is not to say that b_n will never be negative. The subscriber who receives a *prolonged* series of nuisance calls learns not to answer his phone, and in

[15] The curve in Figure 2-2 is drawn on the assumption that the subscriber is risk averse.

[16] In commenting on the first draft of this section, Arthur Okun remarked that this analysis focuses on the benefit or nuisance of an incoming call in isolation, rather than on how the information conveyed by the call interacts with other uncertainties about the world. To the extent that the information conveyed by an incoming call helps to reduce overall risk, then answering the phone is beneficial. Okun notes that it is always bad for his wife when he is going to be home late for dinner, but the bad news is less costly if she learns of it in advance by a call. The bad outcome is made less bad by answering the phone than by not answering it, so that total risk is reduced. Thus it follows that the certainty-equivalent value of an incoming call can exceed the expected value of the call.

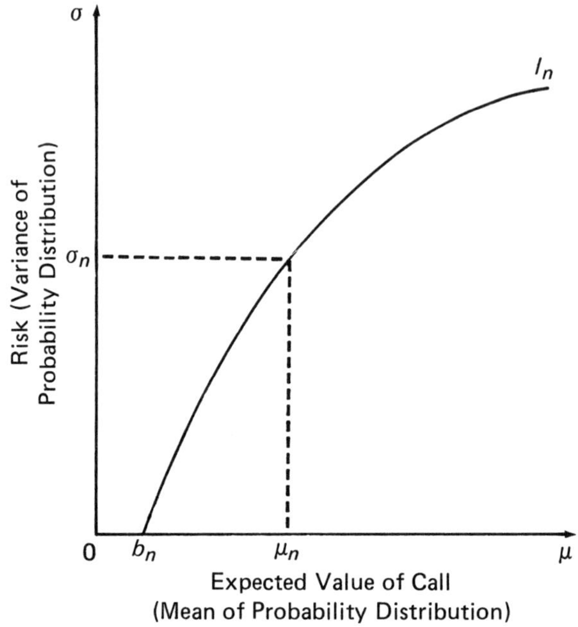

Figure 2-2. Certainty-Equivalent of an Incoming Call for a Risk-Averse Subscriber
Source: Squire (1973), p. 519.

> extreme instances may even have his phone disconnected. Such behavior indicates that $b_n < 0$ for this subscriber. Alternatively, the subscriber may request an unlisted number which may be interpreted as an attempt to insure that b_n is positive, because the unlisted number will only be given to people from whom the subscriber wishes to receive calls. [Squire, (1973, p. 518), italics in original]

As Squire goes on to note, there is not much more that can be deduced about the value of b_n. At present, I know of only two market mechanisms whereby subscribers can express their preferences concerning incoming calls. The first is the firm that offers a toll-free number, while the second is call forwarding, in which the subscriber can have calls to his regular number automatically forwarded to another number. In each case, the value put on incoming calls must be at least as much as the price the subscriber has to pay for the service.[17]

[17] The use of answering service and recording devices are consistent with a positive b_n, but since these services also enable calls to be screened, it would

In the end, Squire assumes that $b_n = b$ for all subscribers. With this assumtion, the total benefits derived from making and receiving calls in the telephone system will be given by

$$TB = \sum_{n=1}^{N} B_n(\overline{N}) + bY$$
$$= \int_0^Z G(z,\overline{N})dz + bY, \tag{36}$$

where Y denotes the total number of incoming calls. For a closed system Y will actually equal Z, since an incoming call is also an outgoing call for someone.

The size of the telephone system is determined in Squire's analysis from the condition that the marginal subscriber will equate the private benefit and cost of belonging to the telephone system. Specifically, for the marginal subscriber, the benefit from making calls $[B_N(\overline{N})]$ plus the anticipated benefit from receiving calls (bY_N) will just equal the cost of making calls (πZ_N) plus the cost of access to the system (r), that is,

$$B_N(\overline{N}) + bY_N = \pi Z_N + r, \tag{37}$$

where Y_N denotes the number of incoming calls anticipated by the marginal subscriber and Z_N denotes the number of calls made by this subscriber.

I have described the analysis of Squire in detail because of the insights that it provides into the nature of access and call externalities and also because of its identification of the dependence of the demand for access on the benefits conferred by use of the telephone system. These are important contributions, and clearly complement the contributions of Artle and Averous and Rohlfs.[18]

---

be incorrect to identify the expenditures for them solely with b_n. As with an unlisted number, their purchase can be viewed as helping to insure that b_n is positive.

[18] Before leaving this discussion, I would like to comment briefly on Squire's use of the conceptual demand function in measuring the benefits from outgoing calls. One way to interpret the conceptual experiment that gives rise to the conceptual demand function for an individual consumer is to assume that income is varied across the other subscribers in such a way as to keep the number of subscribers constant. In this case, the conceptual demand function is an income-compensated demand function, and is accordingly an interesting and useful construct. In particular, it can be used, as Squire suggests (p. 517), to assess a consumer's willingness-to-pay. However, I question that aggregate

A MODEL OF TELEPHONE DEMAND

In this section we shall attempt to mold the ideas and concepts just discussed into a cohesive theoretical structure. Our focus will be strictly on demand, and the question of pricing will be of interest only to the extent that particular pricing schemes have important theoretical and econometric implications. Problems caused by two- and multi-part tariffs are particular cases in point. Similarly, the question of the size of the telephone system will enter into the analysis only to the extent that telephone demand can grow endogenously because of the consumption externalities.

In the discussion that follows, our concern will be with a consumer whose *potential* market basket consists of two goods, telephone calls (q) and a composite good (x) that represents all other goods and services.[19] Let N denote the number of subscribers to the telephone system, and, following Artle and Averous and Rohlfs, define a dichotomous variable such that

$$\delta = \begin{cases} 1 \text{ if the consumer is connected to the} \\ \quad \text{telephone system} \\ 0 \text{ otherwise.} \end{cases} \tag{38}$$

We assume that the consumer has a utility function U with δq, x, and δN as arguments:[20]

$$U = U(\delta q, x, \delta N). \tag{39}$$

willingness-to-pay can be defined, as in expression (33), by summing over individual willingness-to-pay. For individual willingnesses-to-pay cannot be meaningfully summed, as in expression (33), because the individual demand functions are not independent. A particular subscriber's conceptual demand function is derived on the assumption that his income is held constant, while other incomes are varied so as to keep the number of subscribers constant. To sum across individual willingnesses-to-pay thus involves an inconsistency. Partial-equilibrium analysis is being used in a situation that requires a general-equilibrium framework.

[19] The representation of the other goods in the consumer's market basket by a composite good is strictly an assumption of convenience.

[20] We make the standard neoclassical assumptions concerning the properties of U with respect to q and x. In addition, we also assume $\partial U/\partial N > 0$. Also in reviewing the first draft of this chapter, Merton J. Peck suggested that the system externality might be measured better by the proportion of the population having a telephone rather than just the number of subscribers, which is to say that we assume that the system externality is always positive. However, the popularity of unlisted and unpublished numbers in large systems suggests that this may not always be the case. Almost certainly, the effect of the externality is nonlinear.

Finally, let the consumer's budget constraint be given by

$$\delta(r + \pi q) + px = \mu, \tag{40}$$

where

r = price of access to the telphone system
π = price of a call
p = price of the composite good x
μ = income of the consumer.

The utility function in (39) differs from the utility function postulated by Artle and Averous in that the number of calls *and* the number of subscribers appear as arguments, rather than just the number of calls. The number of calls represents the purely private benefits of using the telephone system, while the number subscribers represents the external benefits of system size. Unlike Artle and Averous, we do not assume that each individual communicates with everyone else in the population, and unlike Rohlfs, we do not allow for interdependence among subscribers by including as arguments the calls made by every other subscriber to the telephone system.

Rohlfs' procedure is, in principle, superior to the one that is being followed here because it allows for the call externality as well as the access externality. But the present proceudre is clearly more tractable empirically. Also, as noted by Littlechild (1975), when the consumption of a large group of subscribers is considered—as is the case for most empirical studies of telephone demand—the *individual* effects of call externalities will tend to balance out. For when people have repetitive contacts, each person can return the benefit of receiving a call. However, since the call externalities are not, in general, reflected in either the price of access or the price of use, there will be an unallocated benefit for the *system* as a whole. On the other hand, the size of this benefit, which represents the collective benefits from incoming calls, should be closely related to the size of the system, and (for a large group of subscribers) will thus also tend to be reflected in the number of subscribers.

Let me now turn to the derivation of demand functions for use and access. These will be obtained by postulating, following Rohlfs, a two-step maximization procedure. In the first step, the net benefits from use of the telephone system by an individual are calculated conditional on access, while in the second step, the net benefits are compared with the cost of access in order to determine whether access will in fact be purchaed.

In Step 1, therefore, we assume that $\delta = 1$ and then maximize the LaGrangian function,

$$L = U(q,x,N) - \lambda(\pi q + px - \mu + r) \tag{41}$$

with respect to q and x, where λ is the LaGrangian multiplier associated with the (conditional) budget constraint,

$$\pi q + px = \mu - r. \tag{42}$$

The first-order conditions are accordingly [in addition to (42)]:

$$\frac{\partial L}{\partial q} = U_q - \lambda\pi = 0 \tag{43}$$

$$\frac{\partial L}{\partial x} = U_x - \lambda p = 0. \tag{44}$$

From these three equations, we can in principle derive the demand functions for calls and all other goods:

$$q = q(\pi, p, N, \mu - r) \tag{45}$$

$$x = x(\pi, p, N, \mu - r). \tag{46}$$

These functions differ from conventional demand functions in two respects: (1) the budget constraint is $\mu - r$, rather than μ, which reflects the conditionality of use on having purchased access to the telephone system, and (2) the demand functions depend on the number of subscribers, which reflects the access and call externalities. Determination of the net benefits for the consumer from use of the telephone system proceeds by evaluating the consumer's surplus generated by use of the telephone system. Accordingly, let

$$\pi = g(q, p, N, \mu - r) \tag{47}$$

be the inverse demand function for calls [obtained by solving equation (45) for π]. The consumer's surplus (S) from making q calls will consequently be given by

$$S = \int_0^q g(z, p, N, \mu - r)dz - \pi q. \tag{48}$$

In Step 2, the consumer's net benefits from using the telephone system are compared with the cost of access to the system. If S is at least as great as r, the benefits from belonging to the telephone system outweigh the cost of access, and the consumer will subscribe to the system. Thus,

$$\delta = \begin{cases} 1 \text{ if } S \geqslant r \\ 0 \text{ if } S < r. \end{cases} \tag{49}$$

It is is found in Step 2 that the consumer does subscribe to the system, the conditional analysis in Step 1 becomes unconditional—that is, the demand functions for calls and other goods and services given in expressions (45) and (46) are the actual demand functions. However, if Step 2 shows that the consumer will not subscribe, then equations (45) and (46) do not apply, for $\delta = 0$ also implies that $q = 0$. The utility function in (39) will therefore be given by

$$U = U(x) \tag{50}$$

and the budget constraint will be reduced to

$$px = \mu \cdot \tag{51}$$

The demand function for x will consequently be given by

$$x = \mu/p. \tag{52}$$

The two-step maximization process for determining the demand for use and access combines the procedures of Rohlfs and Squire. Rohlfs, it will be recalled, also employs a two-step maximization framework, but uses the utility function directly to calculate benefits, rather than calculating consumer's surplus as does Squire. In principle, using the utility function to calculate benefits is preferable because it is exact, whereas consumer's surplus provides only an approximation to the benefits. However, as Willig (1976) has shown, the error involved in using a consumer-surplus framework is seldom serious.

The next task is to extend the analysis to an entire population. In the analysis that follows, the demand for use will continue to be identified with the number of calls, while the demand for access will be identified with the proportion of the population that subscribes to the telephone system. Let Q denote the total number of calls that are made by system subscribers.[21] We assume Q to be a function of the price of a call (π), the price of other goods and services (p), the cost of access (r), the number of subscribers (N), and aggregate income (Y):

$$Q = Q(\pi, p, r, N, Y). \tag{53}$$

In arguments, the only difference between the aggregate (or market) demand function and the demand function for an individual subscriber in (45) is that the cost of access (r) is included separately, rather than as a subtration from income. However, we could just as plausibly postulate the budget constraint to be $Y - rN$.[22]

Turning now to the aggregate demand for access, let M denote the size of the population and, as before, let N denote the number of subscribers. In line with (49), define

$$\delta_n = \begin{cases} 1 \text{ if individual } n \text{ subscribes to the} \\ \quad \text{telephone system} \\ 0 \text{ otherwise.} \end{cases} \tag{54}$$

Then,

$$N = \sum_{n=1}^{M} \delta_n . \tag{55}$$

The quantity that we wish to explain is the proportion of the population having access to the telephone system, N/M.

Next, define S_n [as in expression (48)] to represent the consumer's surplus of individual n associated with his use of the telephone system, that is,

$$S_n = \int_0^{q_n} g_n \, (q, p, N, \mu_n - r) dq - \pi q_n . \tag{56}$$

From (49),

$$\delta_n = \begin{cases} 1 \text{ if } S_n \geqslant r \\ 0 \text{ if } S_n < r. \end{cases} \tag{57}$$

---

[21] At this point, all calls are treated as toll calls with π as the price of a call.

[22] Problems of aggregation are being glossed over at this point. Some of these will be discussed in Chapter 5.

Hence, it follows that δ_n is a function of π, p, N, r, and μ_n. Since π, p, N, and r are independent of n, δ_n will vary across individual consumers in the population either because of differences in preferences or because of differences in income. However, for now, we shall assume that everyone has the same preferences.

Since $N = \Sigma\delta_n$ and since δ_n depends on the value of $S_n - r$, N will depend on the distribution of $S_n - r$. However, since preferences are assumed not to vary, it follows that the distribution of $S_n - r$ will depend on the distribution of income. Hence, N, too, will depend on the distribution of income.[23] In particular, for given M, and with S_n treated as a random variable, N will be determined by the probability that S_n is greater than r, namely:

$$P(S_n > r) = 1 - F(r)$$
$$= 1 - \int_0^r f(S_n)dS_n, \tag{58}$$

where $f(S_n)$ and $F(S_n)$ are the density and distribution functions of S_n. Under our assumptions, the distribution of S_n will be related to the distribution of income via the "change-of-variable" from μ_n to S_n as defined in expression (56).

More precisely, let $h(\mu_n)$ represent the density function of μ. Then, employing the formula for a change in variable, $f(S_n)$ will be related to $h(\mu_n)$ according to

$$f(S_n) = h[\mu_n(S_n)] \frac{d\mu_n(S_n)}{dS_n}, \tag{59}$$

where $\mu_n(S_n)$ describes μ_n as a function of S_n, obtained from equation (56) by solving it for μ_n. The other arguments in (56), namely, q_n, p, N, and r, are subsumed into the shape of $\mu_n(S_n)$.[24] Let H denote the distribution function of μ_n. We can now reformulate expression (58) as

$$P(S_n > r) = P[\mu_n > \mu_*(r)]$$
$$= 1 - H[\mu^*(r)]$$
$$= 1 - \int_0^{\mu^*(r)} h(\mu_n)d\mu_n. \tag{60}$$

---

[23] If everyone were also to have the same income, there would be only two equilibrium values that N could take, Q or M. Either no one would belong to the telephone system, or else everyone would belong.

Equivalently, $\mu^*(r)$ can be determined by solving the equation

$$H(\mu^*) = F(r) \tag{61}$$

for μ^* in terms of r.

To summarize: given M and the assumption that tastes do not vary, the proportion of the population that subscribes to the telephone system will be given by:

$$\begin{aligned} \frac{N}{M} &= P(S_n > r) \\ &= P[\mu_n > \mu^*(r)] \\ &= 1 - \int_0^{\mu^*(r)} h(\mu_n) d\mu_n, \end{aligned} \tag{62}$$

or

$$\frac{N}{M} = \int_{\mu^*(r)}^{\infty} h(\mu_n) d\mu_n. \tag{63}$$

Some readers may recognize the expression described by equation (63) as a discrete-choice model, which is a type of model that has been used extensively in recent years in analyzing model choice in transportation demand[25] and fuel and appliance choice in energy demand.[26] In discrete-choice terminology, $\mu^*(r)$ can be interpreted as the "threshold" value of income. If the threshold is exceeded, the consumer subscribes to the telephone system; if the threshold is not reached, the consumer does not subscribe. In the standard formulations of discrete-choice models, the threshold itself is not an object of analysis. Its existence is vital to the model, but how it is determined is not addressed. In the present context, however, we are able to specify how the threshold arises, for it is determined by the consumer's surplus generated by the use of the telephone system compared with the cost of access. Under our assumptions, variation in consumer's surplus across consumers is caused by variation in income. Hence, in this case, the proportion of the population

[24] Since it is assumed that the marginal utility of a call is always positive, S_n in equation (56) is a positive function of μ_n, so that equation (56) is in fact invertible.

[25] See McFadden (1974), Domencich and McFadden (1975), and Hausman and Wise (1978). For a survey of discrete-choice models, see McFadden (1976); see also Nerlove and Press (1973).

[26] See Baughman and Joskow (1975), and Lin, Hirst, and Cohn (1976).

demanding access to the telephone system will be determined by the distribution of income, according to equations (62) and (63).[27]

In view of equations (56), (58), and (59), we can rewrite expression (63) as

$$\frac{N}{M} = \Phi\ (\pi, p, r, N, Y), \tag{64}$$

where Φ is a highly complicated composite function embodying S_n in expression (56), the change-of-variable in equation (59), and the integral in equation (63).[28] The distribution of income is represented in (64) by the variable Y, which (for expositional convenience) we shall identify with the aggregate level of income. In view of the complexity of equation (63), empirical application will probably require a more detailed representation of the income distribution.

Equations (53) and (64) comprise, in general form, a bare-bones model of telephone demand for a population of residential consumers. Equation (53) describes the total number of calls that are logged by subscribers to the telephone network, while equation (64) describes the proportion of the population that subscribes to the network. Equation (53) thus represents the aggregate demand for use of the telephone system, and equation (64) represents the aggregate demand for access to the system. In later sections, we shall extend this model to take into account flat-rate pricing for local calls and toll prices that vary by mileage, duration, time of day, and type of call. A section will also be devoted to business demand.

OPTION DEMAND

In Section 1 it was noted that subscription to the telephone system is closely related to option demand. Option demand was first dis-

---

[27] If tastes also vary, then the proportion of the population demanding access will depend upon the joint distribution of tastes and income. Allowing tastes to vary is mandatory if one is going to deal realistically with situations where consumers can choose among several types of basic service—one-party flat-rate, two-party flat-rate, one-party measured, and so on.

[28] The argument leading up to expression (63) contains a problem similar to the one discussed in connection with Squire's derivation of the aggregate willingness-to-pat function: the number of subscribers, N, is assumed to be fixed in deriving the individual δ_n; however, at the aggregate level, N is variable. As expression (64) stands, it is an implicit function in N, and accordingly can in principle be solved for N. A more practical solution is suggested in Section 9 below.

cussed by Weisbrod in a provocative article in the *Quarterly Journal of Economics* in 1964, and has since figured prominently in the economics of natural resources, particularly in the preservation of irreplaceable places of natural beauty. As described by Krutilla (1967),

> [option] demand is characterized as a willingness to pay for retaining an option to use an area or facility that would be difficult or impossible to replace and for which no close substitute is available. Moreover, such a demand may exist even though there is no current intention to use the area or facility in question and the option may never be exercised. [Krutilla (1967, p. 780)]

In the present context, option demand refers to where the making of a call is contingent upon a state of nature whose realization is uncertain. In this situation, the consumer is willing to pay something for the option to make the call even though the option may not be exercised. Let us assume that, during a given period of time (say a month), a consumer is willing to buy options to make R calls, but that, on the average, only a proportion Θ of these R calls will actually be made.[29] Assume that both Θ and R are known to the consumer. Assume, further, that the expected value of the number of options that will be exercised, ΘR, is included in q (defined in the preceding section), so that the net benefits from these calls are already reflected in the consumer's surplus measure defined in expression (56).

However, the benefits from the $(1 - \Theta)R$ options that are not exercised are not reflected in S_n, which means that S_n *understates* the amount that the consumer is willing to pay in order to have access to the telephone system. Let ω denote the benefit conferred by an option that is not exercised.[30] The benefits associated with the $(1 - \Theta)R$ unexercised options will then be equal to $\omega(1 - \Theta)R$. This, then, is the amount by which S_n will understate the consumer's willingness-to-pay for access to the system.

With option demand in the picture, expression (57) for δ_n becomes

$$\delta_n = \begin{cases} 1 \text{ if } S_n + \omega(1-\Theta)R \geqslant r \\ 0 \text{ if } S_n + \omega(1-\Theta)R < r. \end{cases} \tag{65}$$

---

[29] These R calls are in addition to those the consumer knows will be made with certainty. R may be very large, but for most consumers Θ will be very small.

[30] Since the value of the benefit associated with an option will almost certainly vary with the call involved, ω should be viewed as an expected value.

Expression (45) remains valid as the demand function for use; similarly, equation (53) continues to describe the demand for the total number of calls in the system. However, in deriving the function describing the aggregate demand for access, the threshold value for income in expression (60) must be replaced by $\mu^*(r) - \omega(1-\Theta)R$, which means that expression (62) becomes

$$\frac{N}{M} = \int_{\mu^{**}}^{\infty} h(\mu_n)d\mu_n, \tag{66}$$

where

$$\mu^{**} = \mu^*(r) - \omega\,(1-\Theta)R. \tag{67}$$

Hence, expression (64) must be revised to[31]

$$\frac{N}{M} = \Phi(\pi, p, r, N, Y, \Theta, R). \tag{68}$$

Although the foregoing is straightforward in principle, the important question is whether option demand can be dealt with empirically. Since it is unlikely that information on Θ and R can be found, their presence will have to be inferred indirectly. To this end, let us consider two types of option calls: (1) calls of an emergency nature (police, fire, medical, etc.) and (2) calls of a non-emergency nature related to pleasure or business.

Generally speaking, calls of an emergency nature should be reasonably independent of the size of the local network (or exchange), since small communities, as well as large communities, require doctors and police and fire protection. However, this type of option demand is probably relatively more important in small as opposed to large exchanges and in rural as opposed to urban areas. In a rural area, having access to a telephone is probably more urgent than in an urban area because of the absence of immediate neighbors and because distances are likely to be longer. In a larger exchange, on the other hand, option demand of this type is likely to be of lesser relative importance because the option demand of the second type is likely to be of greater absolute importance.

Calls of the second type will be primarily related to a subscriber's income and his community of interest, both of which are likely to be larger with a larger exchange. Income is higher in a large

---

[31] The value of ω is assumed to be absorbed into the shape of Φ.

exchange, because large exchanges are in urban areas and urban incomes are higher than rural incomes. Consequently, since *R* would seem almost certainly to be heavily weighted with option calls of the second type, network size, in addition to everything else that it might measure, ought also to stand as an indicator of option demand. On the other hand, since option calls of the first type are probably of greater relative importance in communities with low population density, a measure of urbanization ought also be included as a predictor in the equations describing the demand for access.[32]

THE DURATION AND DISTANCE DIMENSIONS OF TELEPHONE DEMAND

For a given customer class, a telephone call possesses four dimensions: type of call, time, distance, and duration. Since the cost of "supplying" a call varies in each of these dimensions, telephone companies have priced the call accordingly. In some cases, the pricing schemes used have been in accord with the principles of marginal-cost pricing (day/night and weekday/weekend differentials, a lower charge for direct dialing, etc.), while in other cases, the motivation has been primarily to subsidize the access and local-service markets by the long-distance market. However, in the present context, the issue is not *why* a telephone call is priced in multi-dimensions, but rather *that* it is so priced. Our task in this section and the next will be to integrate this fact into the theoretical structure.

The first thing that we shall focus on is the complications caused by multi-part tariffs. Most telephone services are priced on some form of multi-part tariff, and accordingly care must be taken in the way that price is introduced into the analysis. A toll call, for example, is supplied on a two-part tariff, the initial period carries one charge, and subsequent periods (or parts thereof) carry a lower

---

[32] Unfortunately, the discussion in this section has barely scratched the surface of option demand and its relation to the demand for access. At this stage, we do not know the quantitative importance of option demand nor how it may vary across customer classes or sizes of exchange. In addition, option demand may be a factor in subscribers' apparent preference for flat-rate pricing for local service, even though measured service may result in a smaller expected overall bill. For some subscribers, anyway, there may be a willingness to pay a premium for having the option of making an unlimited number of calls. This is an area where survey information could be illuminating.

charge. As a consequence, the average price per call-period is no longer the same as the marginal price.

To see what is involved with a multi-part tariff, let us assume that q is supplied according to the following rate schedule:[33]

1st q_1 units of q or less	r $'s
q_1 to q_2 units of q	π_0/unit of q
more than q_2 units of q	π_1/unit of q,

where it is assumed that $\pi_0 > \pi_1$. Assume that a second good x can be purchased in unlimited quantities at price p per unit. As usual, we shall assume that the consumer possesses a utility function $U(q, x)$ that is maximized subject to the level of income μ.

With the multi-part tariff for q, the budget constraint becomes nonlinear. Its general appearance is as given in Figure 2-3, while algebraically it is given by [34]

$$r + \psi_0(\pi_0, q, q_1, q_2) + \psi_1(\pi_1, q, q_2) + px = \mu, \tag{69}$$

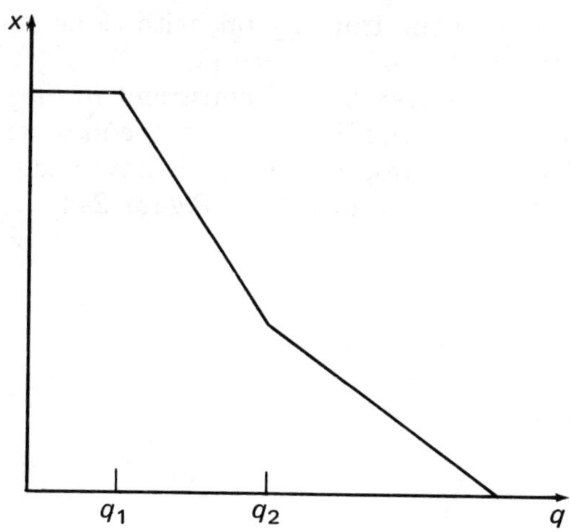

Figure 2-3. Budget Constraint with Decreasing Block Pricing

---

[33] The following discussion is taken from Taylor (1975, pp. 76-79). For the next several paragraphs, q should be viewed as an abstract good that is sold on a multi-part tariff. The point of these paragraphs is to describe the problems caused by multi-part tariffs under highly simplified assumptions. However, if readers prefer at this point to identify q with telephone calls, then q should be viewed as a single call measured in call-minutes (or call-seconds).

[34] It is assumed that the amount r is spent independent of whether any q is actually consumed.

where

$$\psi_0(\pi_0, q, q_1, q_2) = \begin{cases} 0 \text{ if } q \leqslant q_1 \\ \pi_0 (q - q_1) \text{ if } q_1 < q \leqslant q_2 \\ \pi_1 (q_2 - q_1) \text{ if } q > q_2 \end{cases} \tag{70}$$

$$\psi_1(\pi_1, q, q_2) = \begin{cases} 0 \text{ if } q \leqslant q_2 \\ \pi_1 (q - q_2) \text{ if } q > q_2 . \end{cases} \tag{71}$$

The horizontal segment of the budget constraint in Figure 2-3 corresponds to the "fixed charge" of r for consumption of the first q_1 units of q. The linear segment between q_1 and q_2 has slope equal to $-\pi_0/p$, and corresponds to the π_0 block of the tariff schedule. Finally, the segment from q_2 on, with slope equal to $-\pi_1/p$, corresponds to the π_1 part of the schedule.

The nonlinear, nonconvex budget constraint in (69) has a number of consequences for the equilibrium of the consumer, his demand functions, and Engel curves, and we shall now analyze these with the aid of Figures 2-4 through 2-9. Figure 2-4 shows the point

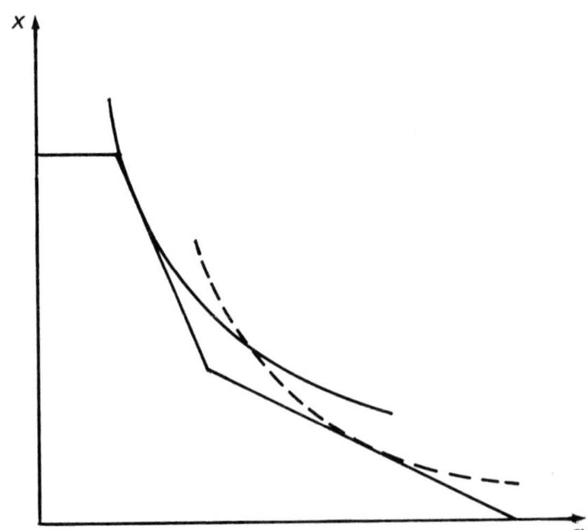

Figure 2-4. Consumer Equilibria with Decreasing Block Pricing

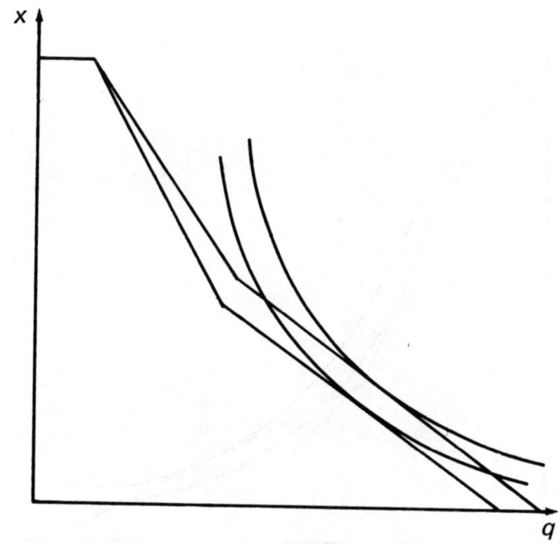

Figure 2-5. Effect of a Change in Intramarginal Price

Figure 2-6. Effect of a Change in Both Intramarginal and Marginal Price

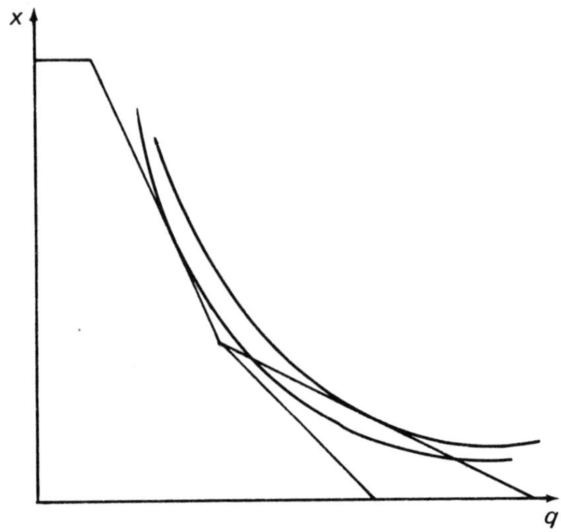

Figure 2-7. Price Change That Leads to Switching of Blocks

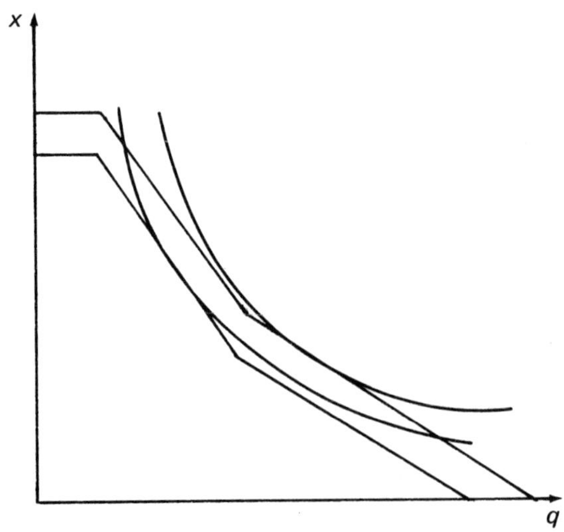

Figure 2--8. Effect of an Income Change

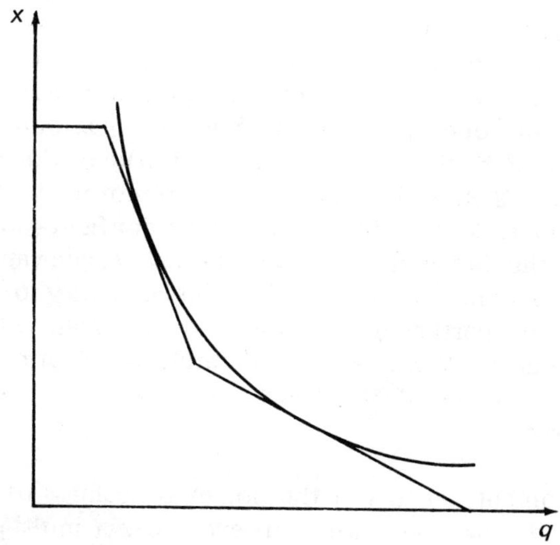

Figure 2-9. Multiple Equilibria

of equilibrium for two different indifference maps. The indifference map corresponding to the solid curve gives an equilibirum on the facet of the budget constraint having slope equal to $-\pi_0/p$, while the one corresponding to the dashed curve gives an equilibirum on the facet with slope equal to $-\pi_1/p$.

Figure 2-5 describes an increase in π_0, but keeps π_1 constant, while Figure 2-6 depicts an increase in π_1 as well. In Figures 2-5 and 2-6, equilibrium following a price increase remains on the same facet of the budget constraint, but it is clear from 2-7 that this need not always be the case, since the price increase in this figure leads the consumer to drop back to a higher rate class. Switching into a different marginal rate class can also come about by a change in income, as is evident in Figure 2-8. Finally, Figure 2-9 shows a case in which the budget constraint is tangent to the same indifference curve at two different points, thereby resulting in multiple equilibria.

Figures 2-4 through 2-9 thus support the following conclusions:

1. Because of the piece-wise linearity of the budget constraint, the equilibrium of the consumer cannot be derived, as is conventionally the case, using the differential calculus. Mathematical programming must be used instead. Demand functions and Engel curves still exist, but they cannot be obtained as closed-

form expressions from solution of the first-order conditions for utility maximization.

2. From Figure 2-6, it is evident that the demand functions are discontinuous, with jumps at the points where equilibrium switches from one facet of the budget constraint to another.
3. From Figure 2-8, the same is seen to be true of the Engel curves.
4. From Figure 2-9, it is evident that for normally shaped indifference curves, there will be particular configurations of prices for which the demand functions are not single-valued. This is, of course, a consequence, of the non-convexity of the budget constraint. In particular, the demand functions will be multi-valued whenever there is a configuration of prices that yield multiple tangencies of the budget constraint to the same indifference curve.

For the moment, let us put the nonanalyticalness of the demand functions to the side and ask oursleves how a multi-part tariff is to be represented in a demand function that is to be estimated empirically. Suppose that the consumer is in equilibrium in the π_1 block, and consider the following three cases:

1. An increase in the "fixed charge" r, π_0 and π_1 remaining the same;
2. An increase in π_0, r and π_1 remaining the same;
3. An increase in π_1, r and π_0 remaining the same.

In Figure 2-10, the increase in r is seen to shift the budget constraint downward, and leads to a reduction in the amount of q consumed. In Figures 2-5 and 2-11, it is seen that q is decreased for Cases 2 and 3 as well. However, it will be noted that the reductions in q in Cases 1 and 2 *arise strictly from an income effect.* That is, an increase in the "fixed charge" or an increase in an intramarginal price are equivalent in the sense that they give rise to income effects, but not to substitution effects.[35] However, an increase in π_1, the marginal price, yields, in traditional fashion, both an income effect and a substitution effect.

From this simple exercise, it is evident that all parts of a multipart tariff up to and including the price in the marginal block belong

[35] The only qualification to this statement arises in the situation where the increase in π_0 or r is sufficiently large that equilibrium switches from the π_1 block to the π_0 block. In this case, there would be a substitution effect as well as an income effect.

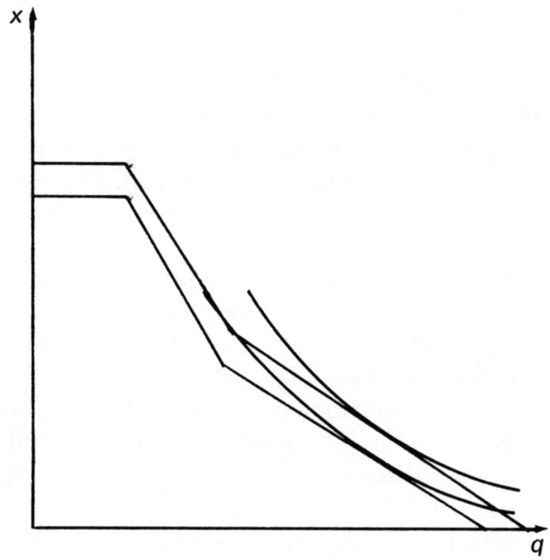

Figure 2-10. Change in Fixed Charge

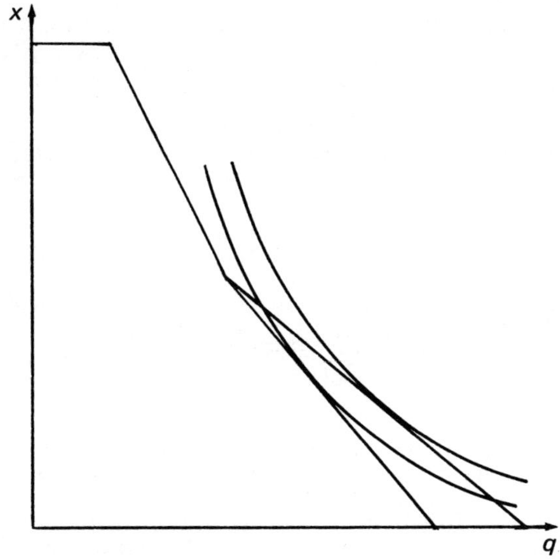

Figure 2-11. Change in Marginal Price

in the empirical demand function. Changes in intramarginal prices (so long as they do not cause the consumer to shift blocks) give rise to income effects only, while a change in the price in the marginal block gives rise to both an income effect and a substitution effect.

Let me now interpret the foregoing as a highly simplified model of telephone demand. One can define r (as before) as the price of access to the telephone system, π_0 as the charge for the initial period of a call, and π_1 as the charge for each overtime period (or part thereof). In this case, then q will represent the duration of a call.[36] The value of q will depend on all three telephone prices (as well as on income and the price of other goods), but the price effect, as we ordinarily think of it, will only be reflected in the charge for an overtime period. Provided that the call lasts into at least one overtime period, the price of access and the initial-period charge will affect duration only through an income effect.[37]

The fact that the demand functions (with a multi-part tariff) are multi-valued and discontinuous for certain configurations of prices means that the demand functions cannot be represented by analytic functions. However, econometric estimation of price and income elasticities requires the use of an analytic function. This necessarily involves an error, but at present we do not know how serious this error may be.[38] Fortunately, though, the problem tends to disappear with aggregation, for as one aggregates over a population of consumers, the average demand functions for the population become continuous as the size of the population becomes large.[39] A sufficient condition for this to occur is that income or tastes vary across consumers. Since most empirical analyses of telephone demand are based on aggregate data, the problems caused by multi-valuedness and discontinuities can thus safely be ignored.

---

[36] With this example, it is best to take q_1 as being equal to zero, q_2 then measures the length of the initial period.

[37] Also, this conclusion ignores the possibility that a person who speaks frequently with another may substitute a large number of short calls for a small number of long calls if the charge for overtime periods becomes large relative to the charge for the initial period. In this case, the initial-period charge would affect duration.

[38] Steven Wade, a graduate student in economics at the University of Arizona, is currently exploring this question, in the context of residential electricity demand, in his Ph.D. dissertation. His initial results suggest that the problem may not be as severe as I had thought.

[39] The theorem involved was proven by Gail R. Blattenberger in her Ph.D. dissertation at the University of Michigan [Blattenberger (1977)]. The theorem is also discussed in Taylor, Blattenberger, and Verleger (1977).

Let me now return to the main purpose of this section, which is to incorporate the duration and distance dimensions of a telephone call into the theory of telephone demand. The discussion that follows will be facilitated by our assuming a specific functional form for the demand functions in (53) and (67). For ease in exposition, we shall postulate these functions to be linear. Hence, for the demand function for use [equation (53)] we assume:

$$Q = \alpha_0 + \alpha_1 \pi + \alpha_2 p + \alpha_3 r + \alpha_4 N + \alpha_5 Y, \tag{72}$$

while for the demand function for access [equation (68)] we assume:[40]

$$m = \beta_0 + \beta_1 \pi + \beta_2 p + \beta_3 r + \beta_4 N + \beta_5 Y, \tag{73}$$

where

$$m = \frac{N}{M}. \tag{74}$$

Let us consider, first, the case where a telephone call is timed and charged according to duration. However, we shall continue to assume that the cost of the call does not vary by type, time-of-day, or length-of-haul. With this manner of charging for calls, the only modification in equation (72) that is required is that Q is now more properly interpreted as referring to the number of call-minutes, rather than to just the number of calls. The price variable, π, in both equation (72) and equation (73) will, of course, refer to the charge measured in some unit of connect time (most likely a minute or fraction thereof). With the redefinition of π, equation (72) does not require further modification.

We shall now complicate the situation (but also make it more realistic) by assuming that a call is priced on a two-part tariff. In particular, assume that the first minute of a call costs π_0, while each subsequent minute, or part thereof, costs π_1, where $\pi_0 > \pi_1$. As we have just seen, a two-part tariff introduces a complication in that the average price per call-minute of a call is no longer the same as the marginal price. There are accordingly two ways that we

---

[40] For convenience, it is assumed that Θ and R, the parameters describing option demand, are reflected in N. Also, we ignore until Section 9 the fact that N appears on both sides of the equation.

can proceed. The first, which follows form the preceding discussion, is to include both π_0 and π_1 in the demand functions as predictors, that is,

$$Q = \alpha_0 + \alpha_1 \pi_0 + \alpha_1' \pi_1 + \alpha_2 p + \alpha_3 r + \alpha_4 N + \alpha_5 Y \tag{75}$$

$$m = \beta_0 + \beta_1 \pi_0 + \beta_1' \pi_1 + \beta_2 p + \beta_3 r + \beta_4 N + \beta_5 Y. \tag{76}$$

In this case, α_1' (assuming that not all calls terminate within the initial period) will measure the traditional price effect (since π_1 is the price on the margin), while α_1 will measure an income effect. Naturally, we expect both α_1' and α_1 to be negative.

Alternatively, equation (72) can be replaced by two equations. The first equation will explain the *number* of calls that are made, while the second equation will explain the *mean duration* of a call. Let Z denote the former, and $\bar{t}$ the latter. Then,

$$Q = Z\bar{t}. \tag{77}$$

The relationship between Z and $\bar{t}$ can be approached in the same framework as the demand for access and the demand for use. Making a call corresponds to the decision to purchase access, while the duration of the call corresponds to use.[41] The first part of the two-part tariff, π_0, represents the "buy in" price for the call, while π_1 represents the "running cost." Thus, π_1 will be most relevant in explaining $\bar{t}$, but both π_0 and π_1 should figure in the equation for Z. The inclusion of π_1 in the equation for Z reflects the fact that, in line with the link between access and use expressed in equation (57), the amount that a subscriber is willing to pay to make a call will depend on the net benefits that the call confers, and this will clearly depend on π_1.

For the equation explaining the number of calls, we therefore postulate:

$$Z = \gamma_0 + \gamma_1 \pi_0 + \gamma_2 \pi_1 + \gamma_3 p + \gamma_4 r + \gamma_5 N + \gamma_6 Y, \tag{78}$$

while for the equation explaining mean duration, we assume:

---

[41] Cf. Khadem (1976).

$$\bar{t} = \kappa_0 + \kappa_1 \pi_0 + \kappa_2 \pi_1 + \kappa_3 p + \kappa_4 N + \kappa_5 Y. \tag{79}$$

Strickly speaking, r should also be an argument in equation (79) since the duration of a call is clearly conditional on access. However, the contribution of this variable will operate only through an income effect, and its net influence on $\bar{t}$ will almost certainly be small.[42] On the other hand, one might question whether system size belongs in equation (79), for, once a call is made, the number of subscribers might seem irrelevant to how long the call lasts. In commenting on the first draft of this section, William Shew of NERA questioned this conclusion. Shew pointed out that if a portion of A's calls to B consist of information that B is to convey to C, who does not have a telephone, then C's getting a telephone would shorten A's conversations with B, since A could now talk to C directly. Shew also noted that if all eight of A's closest friends have telephones, A's average length of conversation is likely to be shorter than if only four of the eight have telephones. Thus, it seems that whether system size affects duration has to be left as an empirical question.

Continuing down the list of complications, we shall now assume that the price of a call varies with distance (i.e., length-of-haul). We shall retain the assumption that a call is charged by the call-minute, but (for now) we shall drop the assumption that this is on a two-part tariff. Further, we shall assume that there are D cities that a subscriber can call. Let these be indexed by j, and assume that they are ordered by the distance that they are from the subscriber. Let π_j denote the price per call-minute of a call to city j and assume that $\pi_j > \pi_k$ for $j > k$. The question that we wish to answer is: What will determine the subscriber's demand for call-minutes to city j?

In approaching this question, it will once again be useful to decompose call-minutes into the product of the number of calls

---

[42] The initial-period charge (π_0) will also affect E through an income effect. However, William Shew of NERA has suggested that there may be a substitution effect as well, because there may be situations where the number of calls and duration can be substitutes. For example, suppose that a person is weighing making three-minute calls to a relative in the middle and at the end of a month or a six-minute call at the end of the month. If π_0 is greater than π_1, then the former would be more expensive, but could be of greater value because the information conveyed is more current. The choice in this situation will depend in part, therefore, on the incentive created by the tariff to substitute call frequency for call duration.

and the mean duration of a call. Hence, parallel to equation (77), we have for calls to city j,[43]

$$q_j = z_j \bar{t}_j. \tag{80}$$

We shall focus first on the determinants of z_j.

As before, it is clear that z_j will depend on prices, income, and some measure of system size. Concerning prices and income, no new questions of principle arise, and these variables can be represented as previously. However, how to represent system size is another matter, for in this case, system size is ambiguous. To this point, it has more or less been implicit that the subscriber's community of interest is geographically local. However, the community of interest has now been extended to another telephone system, and the size of that system will be the relevant measure of size, rather than the size of the local system. Everything else being constant, the larger is city j, the more people in that city the subscriber is likely to know, and, therefore, the more calls that are likely to be made. But distance will also be a factor. In particular, the more distant is city j, the fewer people in city j the subscriber is likely to be acquainted with, and hence the fewer calls that are likely to be made.

For the demand function for z_j, we therefore postulate[44]

$$z_j = \gamma_0 + \gamma_1 \pi_j + \gamma_2 p + \gamma_3 r + \gamma_4 \mu + \gamma_5 N_j + \gamma_6 d_j, \tag{81}$$

where N_j denotes the number of telephones that can be reached in city j and d_j denotes the distance of city j from the subscriber. As noted, we expect γ_5 to be positive and γ_6 to be negative. For the equation for mean duration, $\bar{t}_j$, we assume

$$\bar{t}_j = \kappa_0 + \kappa_1 \pi_j + \kappa_2 p + \kappa_3 \mu + \kappa_4 N_j + \kappa_5 d_j. \tag{82}$$

As in equation (79), the price of access, r, is excluded from this equation, although in principle it belongs. Note, too, that, also

---

[43] Our focus in the next several paragraphs will be on an individual subscriber. This is indicated by the use of lower-case letters, q and z, in place of Q and Z.

[44] The cities other than j that the consumer might call are ignored at this stage. It is not clear a priori how much interdependence there is (if any) among calls to different cities and, besides, it is best not to try to cover everything in the present discussion.

as in equation (79), the number of subscribers in city j is included as an argument. In this case, though, it is not clear that the impact of a larger system size on duration (assuming that system size is important at all) will be negative. For (as noted earlier), the larger is city j, the more people in that city the subscriber is likely to know, so that a call to city j may be longer as a result of more messages being conveyed to acquaintances in city j through the person who is called. Finally, since a positive relationship between the duration of a call and distance is a well-established empirical regularity, we expect κ_5 to be positive.

Let us now turn to the total volume of one-way traffic between the subscribers in one system and the subscribers in another system. Let i and j denote the two systems involved, and let Q_{ij} denote the one-way toll volume, measured in call-minutes, from system i to system j. Let Z_{ij} denote the number of calls from i to j, and let $\bar{t}_{ij}$ denote the mean duration of a call. Then,

$$Q_{ij} = Z_{ij}\bar{t}_{ij}. \tag{83}$$

For the model explaining Z_{ij}, we propose:

$$Z_{ij} = \lambda_0 + \lambda_1\pi_{ij} + \lambda_2 p + \lambda_3 r_i + \lambda_4 N_i + \lambda_5 N_j + \lambda_6 Y_i + \lambda_7 d_{ij}, \tag{84}$$

where

π_{ij} = price/call-minute of a call from system i to system j
p = price of other goods and services
N_i = number of telephone subscribers in system i
N_j = number of telephone subscribers in system j
Y_i = income in system i
d_{ij} = distance between systems i and j.

The variables in equation (84) require little comment. We should, of course, expect λ_1 and λ_3 to be negative and λ_2 and λ_6 to be positive. Since the volume of traffic should vary directly with the sizes of the systems involved, we should also expect λ_4 and λ_5 to be positive. Finally, as in equation (81), we should expect λ_7 to be negative. For the equation explaining the mean duration of a call, we postualte [in line with equation (82)]

$$\bar{t}_{ij} = \xi_0 + \xi_1\pi_{ij} + \xi_2 p + \xi_3 N_j + \xi_4 Y_i + \xi_5 d_{ij}. \tag{85}$$

As in equation (82), the coefficient on n_j may be positive, as should be also the coefficient on d_{ij}.

No new questions of principle arise when (as is in fact the case) a toll call is supplied on a two-part tariff. With slight modification to our previous notation, let π^0 denote the first part of the tariff, and let π^1 denote the second part. In line with equations (78) and (79), equations (79) and (80) for the number of calls and mean duration of a call by a single subscriber to city j become

$$z_j = \gamma_0 + \gamma_1 \pi_j^0 + \gamma_1' \pi_j^1 + \gamma_2 p + \gamma_3 r + \gamma_4 \mu + \gamma_5 N_j + \gamma_6 d_j \tag{86}$$

and

$$\bar{t}_j = \kappa_0 + \kappa_1 \pi_j^0 + \kappa_2 \pi_j^1 + \kappa_3 p + \kappa_4 N_j + \kappa_5 \mu + \kappa_6 d_j. \tag{87}$$

Similarly, equations (84) and (85) describing the total volume of traffic from system i to system j become

$$Z_{ij} = \lambda_0 + \lambda_1 \pi_{ij}^0 + \lambda_1' \pi_{ij}^1 + \lambda_2 p + \lambda_3 r_i + \lambda_4 N_i + \lambda_5 N_j + \lambda_6 Y_i + \lambda_7 d_{ij} \tag{88}$$

$$\bar{t}_{ij} = \xi_0 + \xi_1 \pi_{ij}^0 + \xi_1^1 \pi_{ij}^1 + \xi_2 p + \xi_3 N_j + \xi_4 Y_i + \xi_5 d_{ij}. \tag{89}$$

TIME-OF-DAY PRICING

In an effort to smooth out peak-load traffic, telephone companies have for years priced toll calls differently depending on the time of the day or day of the week that a call is placed. We shall now consider the implications of this type of pricing, under the assumption that there are two pricing periods, day and night. Let the price of a call be π^d during the day and π^n at night, and let q^d and q^n denote the number of day and night calls respectively. The utility function in expression (39) will therefore become

$$U = U(\delta q^d, \delta q^n, x, \delta N), \tag{90}$$

and the budget constraint becomes

$$\delta(\pi^d q^d + \pi^n q^n + r) + px = \mu. \tag{91}$$

The demand functions for q^d and q^n *conditional* on access [cf. expression (45)] will be given by

$$q^d = q^d\ (\pi^d, \pi^n, p, n, \mu - r) \tag{92}$$

$$q^n = q^n\ (\pi^d, \pi^n, p, N, \mu - r). \tag{93}$$

Whether or not access is purchased will be determined by the consumer's surplus generated by *both* day calls and night calls in comparison with the cost of access, r. That is, access will be purchased (i.e., δ will be equal to 1) if

$$S_d + S_n \leqslant r, \tag{94}$$

where

$$S_d = \int_0^{q_d} g^d\ (z, \pi^n, p, N, \mu - r)\, dz - \pi^d q^d \tag{95}$$

$$S_n = \int_0^{q_n} g^n\ (z, \pi^d, p, N, \mu - r)\, dz - \pi^n q^n \tag{96}$$

and where $g^d(\cdot)$ and $g^n(\cdot)$ are the inverse demand functions from (92) and (93).

From the foregoing, it should be clear that the introduction of a day/night price differential does not raise any new questions of principle. There will be a demand function for day calls and a demand function for night calls, and π^d and π^n will be arguments in both functions. In general, we should expect day and night calls to be substitutes, which is to say that $\partial q^d/\partial \pi^n$ and $\partial q^n/\partial \pi^d$ should both be expected to be positive.

We shall illustrate the modifications required by time-of-day tariffs using the equations describing the volume of one-way toll traffic between two systems discussed at the end of the preceding section. That is, we shall assume that a call is charged according to time (day/night), distance, and duration and, further, that duration is charged on a two-part tariff. For the equations explaining the number of calls, we will have [cf. equation (87)]

$$Z^d_{ij} = \lambda_0 + \lambda_1 \pi^{0d}_{ij} + \lambda_2 \pi^{1d}_{ij} + \lambda_3 \pi^{0n}_{ij} + \lambda_4 \pi^{1n}_{ij} + \lambda_5 r_i + \lambda_6 p + \lambda_7 N_i + \lambda_8 N_j + \lambda_9 Y_i + \lambda_{10} d_{ij} \tag{97}$$

$$Z_{ij}^{n} = \lambda_0' + \lambda_1' \pi_{ij}^{0d} + \lambda_2' \pi_{ij}^{1d} + \lambda_3' \pi_{ij}^{0n} + \lambda_4' \pi_{ij}^{1n} + \lambda_5' r_i + \lambda_6' p + \lambda_7' N_i + \lambda_8' N_j + \lambda_9' Y_i + \lambda' d_{ij}, \quad (98)$$

while for the equations explaining the mean duration of a call, we will have [cf. equation (89)]:

$$\bar{t}_{ij}^{d} = \xi_0 + \xi_1 \pi_{ij}^{0d} + \xi_2 \pi_{ij}^{1d} + \xi_3 \pi_{ij}^{1n} + \xi_4 p + \xi_5 N_j + \xi_6 Y_i + \xi_7 d_{ij} \quad (99)$$

$$\bar{t}_{ij}^{n} = \xi_0' + \xi_1' \pi_{ij}^{0d} + \xi_2' \pi_{ij}^{1d} + \xi_3' \pi_{ij}^{1n} + \xi_4' p + \xi_5' N_j + \xi_6' Y_i + \xi_7' d_{ij}. \quad (100)$$

OPERATOR-HANDLED VERSUS DIRECT-DIALED CALLS

When a subscriber wishes to make a toll call, there is a choice of dialing the call directly or using an operator. Since an operator-handled call necessarily involves the time of an operator, telephone companies charge more for this type of call, but an operator-handled call also provides more service in that the call can be sent person-to-person, collect, or billed to a third party. This section provides a brief discussion of the factors that enter into the choice of operator-handled versus nonoperator-handled calls. Most of the focus will be on station-to-station versus person-to-person calls. Let π^p denote the price of a person-to-person call, and let π^s denote the price of a station-to-station call.[45] Let η denote the probability that the person the subscriber wishes to talk with can in fact be reached.[46] Then, given a choice between a person-to-person or a station-to-station call, the subscriber will choose the station call if

$$\pi^s \leqslant \eta \pi^p, \quad (101)$$

[45] If a call is charged on a two-part tariff, π^p and π^s are assumed to refer to the first part of the tariff. Current rate schedules typically distinguish between person-to-person and station-to-station calls only in the charge for the initial period. Overtime periods are charged the same.

[46] The probability η is defined on the assumption that the call is completed. If there is no answer or the line is busy, the caller does not pay. The analysis that follows assumes that the call is completed.

that is, if the price of the station-to-station call is no greater than the probability that the person that it is desired to reach is in fact reachable times the price of the person-to-person call. To illustrate, if the probability of reaching the desired party is 0.5, then the caller will be better off making the call station-to-station if the price of the station-to-station call is no greater than one-half of the price of the person-to-person call.[47]

Turning now to the toll calls of a system of subscribers, let τ denote the proportion of the toll calls from the system that are station-to-station calls. We assume that, in the aggregate, this proportion will depend on the *ratio* of π^s to π^p. In particular, we shall postulate

$$\tau = \omega_0 \left(\frac{\pi^s}{\pi^p} \right)^{\omega_1}. \tag{102}$$

In this expression, the effect of τ of the probability of callers reaching desired callees is represented in the parameters ω_0 and ω_1.[48]

The other two types of operator-handled calls, collect calls and third-party billings, should be closely related to travel. This should especially be the case for third-party billings. Collect calls, on the other hand, should be particularly sensitive to college enrollments and military manuevers, but are probably less sensitive than third-party billings to changes in income and price. The income elasticity for credit-card calls, in particular, is probably substantial.

BUSINESS DEMAND FOR TELEPHONE SERVICES

Despite its importance for telephone industry revenues, business demand has received very little attention. The literature includes

---

[47] The decision rule expressed in expression (101) can be complicated for a business call by the likelihood that the caller will attempt to shift part of the cost of the call onto the callee when the latter cannot be reached by asking that the call be returned. In this case, a caller may choose a station-to-station call even when the probability of reaching the person desired is zero. Also, it should be noted that this analysis ignores the fact that the charge for person-to-person call does not begin until the desired party is actually on the line. In some cases, therefore, a person-to-person call may be cheaper than a station-to-station call.

[48] The probability of the desired party being reached will almost certainly vary from call to call, which means that, for the system as a whole, π in expression (102) will be described by a probability distribution. Provided that this distribution is independent of π^s/π^p (which should be the case), its parameters can be represented in the parameters ω_0 and ω_1.

a few empirical studies, but essentially nothing theoretical. There are several reasons for this lack of attention. In the first place, the service purchased by business customers is much more complex than that purchased by residential customers. Businesses can choose from a wide array of terminal equipment and from among message toll service (MTS), WATS, and private line for network services.[49] Hence, right from the start, business demand is more complex to model than residential demand. In the second place, business demand is derived from decisions about what and how much to produce, so that theoretical guidance must be provided by the theory of the firm, rather than the theory of consumer choice. However, most empirical demand analyses are approached from the theory of consumer choice, and, as a consequence, the theoretical underpinnings of business demand have never been of much interest.

The heterogeneity of businesses is a third contributing factor. Businesses vary from small retail establishments to giant multinational corporations, and the variation in telecommunications needs is obviously enormous. Households also vary, but the diversity in tastes across households is probably easier to model than the diversity in the structure of production across businesses. Thus, there is a greater need for disaggregation in modeling business demand. However, disaggregation, while always popular in the abstract, is much less popular in practice because the data needs are usually substantial. Businesses typically have a number of locations, so that billing addresses and locations of actual telephone use may not coincide, and complex corporate structures may make it difficult to identify the industries that the business participates in. A final factor contributing to the tendency to ignore business demand is the fact that, in the past, the telephone companies have never had to use very sophisticated models in justifying proposed changes in business tariffs. As a consequence, there has been no particular pressure to develop specific models of business demand, as opposed to telephone demand in general.

However, in sharp contrast with the past, there is a great deal of current interest in business demand.[50] Indeed, one of the major criticisms of the first draft of this monograph was that I had es-

---

[49] Private line service can also be an option for local calls.

[50] The reasons for this are readily apparent: The business market figures prominently in the current debate in the communications and computer industries (e.g., the introduction of specialized common carriers into the intercity market and new vendors into the terminal equipment market fostered by FCC decisions, private antitrust suits as well as the suits brought by the Department of Justice against AT&T and IBM, and the current effort in the Congress to rewrite the Communications Act of 1934).

sentially ignored business demand. I had not given this much thought at the time, but in retrospect the criticism was clearly on target. Business demand was given short shrift, but as the preceding paragraphs have tried to make clear, the criticism also applies to the existing literature as a whole. In the remainder of this section, I shall indicate some of the factors that will need to be taken into account in developing a theory and model of business demand. Unfortunately, the discussion that follows does little more than touch on high points.

As noted above, a major difference between business and residential telephone demand is that business demand is a derived demand. Firms do not demand telephone service as an end in itself, but as a means to an end, the end (presumably) being the maximization of profits. Profit maximization thus replaces utility maximization as the behavioral hypothesis underlying derivation of the telephone demand function. Telephone service is an input into the production process, and telephone demand is accordingly an input demand. The parameters of the demand function will therefore depend upon the parameters of a firm's production function. However, since there is tremendous diversity among firms in terms of size, output, and production processes, a certain amount of disaggregation seems mandatory. But it is doubtful whether existing data bases will support even the most cursory disaggregation. At a minimum, it would seem that business customers should be disaggregated to retail and wholesale trade, financial institutions, transportation, manufacturing, and government. While there would still remain a great deal of heterogeneity within groups, especially between large and small establishments, this classification would be a useful first step.

The role of telecommunications in the production process is a very important issue, but, basic as this question is, we really do not know much about it. What is clear, though, is that, whatever this role, it will vary from business to business. This is a major reason for wishing to disaggregate. For some businesses, the telephone is a major sales instrument, while for other businesses its primary use is for intrafirm communication. The alternative to the telephone in the first case may be a salesperson in the field, while in the second case, the alternative is the mail or a courier. In both cases—but especially the second—the opportunity cost of time will be a major factor.[51]

---

[51] The opportunity cost of time is a neglected area in telephone demand, not only in business demand, but in residential demand as well. I am aware of only one study, namely, Beauvais (1977), that attempts to incorporate the value of time into a telephone demand model.

Even if the relevant substitutes for telephone communication can be identified, their measurement may be difficult or even impossible. For example, when the alternative to the telephone is a salesperson in the field, the costs involved are the cost of the calls and the travel expenses of the salesperson (and possible his salary). However, face-to-face contact may result in higher sales, so differentials in productivity between the alternatives will also have to be taken into account. But the data needed to measure the costs and benefits involved may not be available.

On the other hand, one should not conclude that business telephone demand has nothing in common with residential demand. The basic character of the telephone system remains. There is still a distinction between access and use, system and call externalities continue to exist, and option demand contributes to the demand for access. Quite clearly, though, there is a shift in the relative importance of these factors between residential and business customers. The access decision for business customers is more a question of the type of terminal equipment to be obtained rather than whether access itself is to be purchased. The type of terminal equipment that is chosen (PBX, centrex, etc.) will depend upon the expected benefits from using each type of equipment weighed against the equipment's cost, with the one chosen being the one that makes the greatest contribution to profits.

The importance of the consumption externalities will almost certainly vary from business to business, and this provides yet another reason for disaggregation. Retail establishments should be particularly cognizant of the system externality, because (at least for moderate-sized systems) another telephone is another potential customer. For large companies, however, the system externality is probably of lesser importance because of the greater frequency of intra-company calls. As regards the call externality, the value of incoming calls for many businesses is so great that the calls are made free to the callers. But there are other businesses for which the value of incoming calls may be of relatively little importance. In this case, the importance of the externality will depend more on the nature of the firm's output than on its size.

One of the biggest complications in modeling business toll demand arises from the fact that many businesses subscribe to (outward) WATS and private line in addition to regular toll service. As a consequence, for many long-distance calls, a business has the option to send the call by private line, WATS, or MTS. Indeed, in many cases (depending upon the terminal equipment available), the call will be sent by a combination of private line and WATS, private line and regular MTS, or even a combination of all three. When the value of time is taken into account, the selection of the least-cost mode of sending the call can become complicated and time-consuming itself. As a consequence, devices have recently become available—the so-called "smart" switches—that place a call automatically by the least-cost mode. Among other things, the value of various employees' time to the firm can be explicitly taken into account in the programming of the "smart" switch. The advent of such devices has at least two important implications for the modeling of business demand: (1) it emphasizes the importance of allowing for technilogical change, and (2) it clearly creates new demands on modeling ingenuity!

SOME FURTHER DYNAMICS

The purpose of this section is to introduce some dynamic elements into the theory, in addition to those inherent in the access/use distinction, by allowing for inertia (or habit formation) in both the access and use equations. In order to keep the analysis simple, we shall revert to our very first price assumption, namely, that there is just one price π that applies to all calls. Our focus in the discussion that follows will be on equations (45) and (49) for an individual subscriber and equations (53) and (63) for a full system of subscribers.

We shall begin with an individual subscriber's demand for telephone calls. Let us assume that when there is a change in price or income, a subscriber does not adjust immediately to a new equilibrium number of calls, but that adjustment occurs over a period of time. Let q_t^* denote the desired (i.e., equilibrium) number of calls during period t for given prices, income, and the number of subscribers. In particular, assume

$$q_t^* = \alpha_0 + \alpha_1 \pi_t + \alpha_2 p_t + \alpha_3 N_t + \alpha_4 (\mu_t - r_t). \tag{104}$$

Next, let q_t denote the actual number of calls that the subscriber makes during the period. Finally, we postulate that whenever q and q^* diverge, a proportion ϕ of the discrepancy is eliminated within any one period. That is, we assume that

$$q_t - q_{t-1} = \phi \, (q_t^* - q_{t-1}), \tag{105}$$

where $0 < \phi \leqslant 1$. Substituting expression (104) into (105) for q_t^*, we obtain (after some rearrangement)

$$q_t = \alpha_0\phi + (1 - \phi)q_{t-1} + \alpha_1\phi\,\pi_t + \alpha_2\phi p_t + \alpha_3\phi N_t + \alpha_4\phi(\mu_t - r_t). \tag{106}$$

Readers will recognize this expression as the estimating equation for the Koyck geometric distributed-lag model, although in this case it has been derived from a Stone-Nerlove partial-adjustment mechanism.

The dynamics that are implicit in expression (106) are simple and easy to derive. Assuming that ϕ is less than 1, the underlying dynamical process is stable, and long-run equilibrium will be achieved when $q_t = q_{t-1}$, or when $q_t = q^*$. Hence, when there is a change in price (say), the long-run impact of this change on q will be given by α_1 [cf. expression (104)]. However, this will be achieved only after an infinity of periods have passed. In the short run, the impact on q will be tempered by the parameter ϕ. Indeed, one way that we can define the short run is as the elaspse of one period following a change in price (or income). From equation (106) we see that the one-period effect on q of a one unit change in price will be equal to $\alpha_1\phi$. Thus the short- and long-run derivatives of q with respect to price are equal to $\alpha_1\phi$ and α_1, respectively. Similarly, the short- and long-run derivatives of q with respect to income are $\alpha_4\phi$ and α_4, and so on for the derivatives with respect to the other predictors.

As before, the demand function in (106) has to be interpreted as conditional on access to the telephone network. However, since the demand for access depends on the consumer's surplus generated by use, it is clear that, unless we assume otherwise, the demand for access will be affected by the inertia in use. At one extreme, we could assume that the consumer's surplus from use (S) is calculated using the demand function in (106). In this case, the access/no access decision would be reopened each period within the adjustment interval. The other extreme would be to assume that S is calculated using the long-run equilibrium value of q (q^*), in which case the demand for access would be insulated completely from the inertia in use.

The problem with the first assumption is that, for a marginal subscriber, $S(q_t)$ [i.e., the consumer's surplus calculated on the basis of q_t in expression (106)] could be less than r_t for some t within the adjustment interval even though $S(q^*)$ is greater than r_t. If the subscriber were already connected to the system, he would not subscribe during the time that $S(q_t)$ were less than r_t, but would subscribe as q_t approaches q^*. Such stop-go behavior seems highly implausible empirically, and we shall therefore assume that S is calculated using q^*.[52]

Let us now return to the full system of subscribers. For the aggregate demand for calls, we shall postulate a partial-adjustment mechanism parallel to the one in (105):

$$Q_t - Q_{t-1} = \psi(Q_t^* - Q_{t-1}). \tag{107}$$

Similarly, we shall assume for Q_t^*:

$$Q_t^* = \alpha_0 + \alpha_1 \pi + \alpha_2 p + \alpha_3 r + \alpha_4 N + \alpha_5 Y. \tag{108}$$

Consequently,

$$Q_t = \alpha_0 \psi + (1 - \psi)Q_{t-1} + \alpha_1 \psi \pi_t + \alpha_2 \psi p_t + \alpha_3 \psi r_t + \alpha_4 \psi N_t + \alpha_5 \psi Y_t. \tag{109}$$

Equation (109) has the same form as equation (106) and therefore has the same dynamical properties.

To be consistent with the behavior postulated for an individual subscriber, the inertia assumed in the aggregate call market cannot affect the aggregate demand for access. However, this does not mean that the aggregate market for access cannot be assumed to be dynamic in its own right. Such an assumption is plausible per se and also provides a solution to the problem that was noted earlier in deriving the aggregate demand for access. This problem, it will be recalled, involved the proportion of a population subscribing to the telephone system itself being a function of the number of subscribers.[53]

Let m^* denote the equilibrium proportion of a population that purchases access to the telephone system for given values of the population, prices, and the distribution of income. In view of expression (64), we will have

$$m^* = \phi(\pi, p, r, N^*, Y), \tag{110}$$

[52] Insulating the demand for access from short-run inertia in use is consistent with the view that the demand for access involves primarily long-run considerations, while the demand for calls is much more sensitive to short-run phenomena. Also, service-connection charges, which have been ignored throughout this chapter, are an important factor in diminishing stop-go connection/disconnection behavior.

[53] See footnote 28.

where N^* denotes the equilibrium number of subscribers corresponding to m^*.[54] Let $m_t = N_t/M$ denote the observed value of m during period t. We now assume that m_t is determined according to[55]

$$m_t = \beta_0 + \beta_1 \pi_t + \beta_2 p_t + \beta_3 r_t + \beta_4 N_{t-1} + \beta_5 Y_t. \tag{111}$$

However, stability requires that β_4 be smaller than $1/M$. If this condition holds, long-run equilibrium will be achieved when $N_t = N_{t-1} = N^*$. Hence, we will have

$$m^* = \beta_0 + \beta_1 \pi + \beta_2 p + \beta_3 r + \beta_4 N^* + \beta_5 Y. \tag{112}$$

The equilibrium number of subscribers can then be obtained by solving this expression for N^*.

CONCLUDING COMMENTS

The primary purpose of this chapter has been to make a start on developing a theoretical framework for analyzing the demand for telephone services. To this end, a collection of demand models has been analyzed that reflects the peculiar features of the telephone. The discussion has emphasized the distinction between access and use, the call and access externalities, and option demand. The problems caused by pricing telephone services on multi-part tariffs have also been emphasized. However, the discussion in this chapter, the number of pages notwithstanding, provides little more than a point of departure. While each of the models discussed in Sections 5, 6, 7, and 9 have been specified with empirical application in mind, the primary motivation has been the illustration of ideas and concepts, and, as a consequence, the models are, for the most part, ad hoc and highly simplified. Linearity, for example, has been assumed throughout, but this has been strictly an assumption of convenience. Nonlinear functional forms are perfectly feasible, and will probably frequently yield superior empirical results.

Finally, I want to warn readers not to take the models that have been presented too literally. Telephone demand is a complex subject, and there is much that is not known. Business demand

---

[54] N^* corresponds to an equilibrium user set in the sense of Rohlfs (1974).
[55] Cf. expression (68).

stands out in this regard, and an entire monograph could easily be devoted to its study. I have emphasized repeatedly the importance of the call and access externalities, option demand, and communities of interest, but how are these concepts to be measured? I have tended to roll both the externalities and option demand into the size of the system (as measured by the number of subscribers), but this can hardly be considered a satisfactory solution, since it leaves the individual contributions underidentified. Also, much remains to be done in incorporating the joint distribution of tastes and income into the aggregate demand for access. Equation (63), as has been noted, is a highly conplicated composite function, and giving it a meaningful specific form will clearly present some challenge. I could go on, but it is time to turn to the empirical literature.

Chapter 3

The Empirical Literature on Telephone Demand: Access, Local Service, and Interstate Toll

In the next two chapters, we undertake a comprehensive review of the many and varied attempts to quantify the behavioral parameters of telephone demand. As we saw in Chapter 1, the empirical literature (when unpublished memoranda are included) is too large for an exhaustive individual review to be feasible. Accordingly, my procedure will be first to summarize the empirical evidence on price and income elasticities of demand, and then to provide detailed reviews of only a few selected studies. In doing this, my object has been to provide a representative picture of the existing state of the art.

The organization of these two chapters is as follows: we begin in Chapter 3, Section 1, with a briew description of the supply side of the telephone industry and a brief summary of the services that are offered to customers. Chapter 3, Section 2-4, and Chapter 4, Sections 1-5, summarize models and results for domestic demand, and Section 6 does the same for international demand. Section 7 is a digression in that it focuses on the demand for yellow-pages advertising. Finally, Section 8 provides a very brief survey of non-econometric approaches to analyzing telephone demand.

SERVICES OFFERED BY THE TELEPHONE COMPANIES

Although our focus in this study is on demand, a brief consideration of the supply side of the telephone industry will be useful in understanding the categories of the telephone service that are offered to the public. In general, telephone companies provide the facilities for collecting, switching, and distributing voice, data, and video messages. The equipment and facilities required consists of[1]

1. Telephones (and related customer-premise terminal equipment);
2. Local loops (i.e., wires between subscribers' premises and the central office);
3. Central offices, which switch and interconnect lines and trunks;
4. Wire centers, a building location housing one or more central offices;
5. Toll central offices, which switch intercity (or interarea) calls;
6. Trunking circuits, which connect two switching systems.

Telephone companies have traditionally charged for access to the telephone system and its use as follows:[2]

1. A service-connection charge when a subscriber is first connected to the system;[3]
2. A monthly local-service charge that is associated with the joint and common costs of the telephone system access and, in most cases, some (or all of) local usage;
3. A fee for toll calls that varies according to a call's type, distance, duration, time-of-day, and day-of-week;
4. Service-connection and local-service charges that are higher for business subscribers than for residential subscribers.

Local service/toll and business/residence splits define the primary dimensions along which telephone companies have marketed their services, but these are by no means the whole story. For telephone companies offer local and toll services in a rather wide variety of forms—measured service versus flat-rate, single-line versus multiparty, WATS versus message toll, to mention a few of the most familiar—and, in some instances, the distinction between local and toll service is severely blurred.[4] Besides basic and toll services, the

---

[1] We ignore the equipment required by international calls. Readers interested in a nontechnical discussion of telecommunications engineering are referred to Bell Telephone Laboratories (1977).

[2] The equipment associated with access includes the main telephone set (and other terminal equipment), inside wiring, the local loop of wires connecting the residence or business, and any nontraffic sensitive dedicated switching equipment.

[3] The telephone companies usually refer to a charge of this type as a *nonrecurring* charge.

[4] The introduction of extended area service (EAS), for example, means that some calls that were previously toll calls become included in local service,

telephone companies provide a variety of other services, including the provision of optional terminal equipment (the so-called vertical services), private line, coin stations, international service, and yellow-pages advertising.

THE DEMAND FOR ACCESS

Despite its importance to an understanding of telephone demand, the demand for access has not received much empirical attention. I have found only eight studies—Alleman (1977), Feldman (1976), Davis et al. (1973), Perl (1978), Pousette (1976), Rash (1971), Southern New England Telephone Co. (1977a), and Waverman (1974)—that contain equations for the demand for access. Alleman, Perl, Rash, and SNET focus on the residential market, while Feldman, Pousette, and Waverman estimate equations for residence and business telephones separately. Pousette also has an equation for the telephones in the public sector, and, finally Davis et al. estimate an equation for the total number of main stations. The focus is the United States as a whole for Alleman, Davis et al., Feldman and Perl; Connecticut for SNET; Canada for Rash; Canada and Sweden for Waverman; and Sweden for Pousette. Davis et al., Pousette, Rash, SNET, and Waverman all employ aggregate time-series data, while Alleman, Feldman, and Perl use cross-section data. The observations are cities for Alleman, states for Feldman, and households for Perl.

The Perl study provides by far the most intensive study of access demand in the literature. Data for individual households from the 1970 Census are used to analyze the economics and demographic determinants of telephone availability in the residential sector. Perl's study has a number of interesting results, and will be described in detail below. Alleman's analysis is also quite intensive, but the focus is on the pricing of local service, rather than on access per se. Alleman devotes an entire chapter in his study (which was his Ph.D. dissertation at the University of Colorado)

---

while with the advent of measured service, local calls that were previously included in flat-rate can become subject to a charge. In the access/use framework, a call is a call whether it be local or toll, although the demand function for "local" calls may differ from the demand function for "long-distance" calls. That a local call has come to be viewed in the literature as something intrinsically different from a long-distance call is an aritificial distinction arising from the way that telephone companies and regulators have chosen to price basic service. Consequently, the fact that the distinction between local and toll service becomes blurred with EAS and measured service is hardly surprising.

to the theory of telephone demand. Following Squire (1973), Alleman relates the demand for access, in a consumer's surplus framework, to the demand for use. Alleman's analysis, however, is disappointing, for the theoretical structure that is carefully built up in the early chapters of the study is ignored almost entirely in the empirical chapters.

The Feldman study is the most comprehensive analysis of telephone demand in the literature. Thirty-six categories of service provided by the Bell System are analyzed, including residence and business main stations and business WATS lines (both inward and outward). As noted above, the analysis is cross-sectional, with states (the Bell System parts thereof) as observations. In the equation for main stations in the residential sector, main stations are assumed to be a function of the price of basic service (measured by average local-service revenues per main station), Bell area population, per-capita personal income, and a set of regional dummy variables that allows the constant term to vary among regions. The population is intended as a measure of market size, but population will also be a proxy for system size, and its coefficient will therefore reflect the access externality. The equation for business main stations is similar, except that local revenue per main station (as a measure of the price of access) is omitted.

A major shortcoming of the Feldman study is the absence of a well-defined theoretical framework. An excellent data base was constructed, and, as we shall see throughout these two chapters, a number of interesting and useful results have been obtained. However, the results lack the crispness and precision of interpretation that a well-articulated theoretical structure would have provided. The study's methodology clearly distinguishes between access and use (since access and use are modeled separately), but one is left with the impression that the dependence of access on the benefits generated by use and the conditionality of use on access are not fully appreciated. Another shortcoming of the study is in the measures of price that are used. Average revenue per unit (of whatever that is being explained) is used for the price variable, but this can lead to problems that are potentially very serious.

In the equations for main stations, for example, defining the price of access as local revenues divided by the number of main stations has three problems: (1) With main stations as both the dependent variable and in the donominator of the price variable, a negative relationship between the number of main stations and the "price" of access is necessarily established; (2) variations in service quality (single-party versus two-party, etc.) will be reflected in local revenues, and these, in turn, will be reflected in average revenue per main station; (3) in areas with measured service, average revenue per main station will reflect charges for local use, as well as the price of access. In this case—as indeed in all cases—the measure of price that is used should only record *shifts* in the underlying tariff schedule, not *movements along it.*[5]

Let me now take up a more detailed discussion of the study of Perl. The questions addressed in Perl's study can be summarized as follows:

1. How is the demand for basic telephone service influenced by characteristics of the household?
2. How is the demand for basic telephone service influenced by the price charged for telephone service?
3. Does the price elasticity of demand vary with income and age?

The determinants of telephone availability are explored using data contained in the household and individual records of the 1/1000 Public Use Sample from the 1970 Census. The sample actually used contains data on 36,703 households in the United States in 1970. In the 1970 Census, respondents were asked whether or not there was "a telephone on which people in your living quarters can be called." The answer to this question, coded as a dichotomous, (0,1) variable, defines access (availability) in Perl's study.[6] Three forms of models are estimated, linear, logit, and probit. In the linear model, the probability of a household having access to a telephone conditional on a particular set of economic and demographic characteristics is assumed to be a linear function of those characteristics. Let x denote vector of economic and demographic characteristics and let $P(T|x)$ denote the probability that a househole has access to a telephone conditional on x. Then, in the linear model,

$$P(T|x) = x\beta + \epsilon, \tag{1}$$

[5] I do not mean to single out Feldman's study as the only transgressor in the use of average revenue as the price variable. Many other studies do it—including Dobell, Taylor et al. (1972)! For a more detailed discussion of the econometric problems involved, see Taylor (1975).

[6] Since some households which do not subscribe to telephone service may nevertheless have access to a telephone (at a nearby coin station or a neighbor's telephone) and some which do subscribe may indicate otherwise (because the number is unlisted, for example), a positive response to the Census question unfortunately cannot be precisely equated with subscription.

where β is a vector of coefficients and ϵ is a random error term.

The virtue of the linear model is that it can be estimated conveniently and inexpensively as a linear regression model. However, it has the drawback that the "conditional probabilities" estimated are not necessarily bounded between 0 and 1.[7] The logit and probit models do not have this defect but they are also much more expensive to estimate.[8] The logit model takes the form

$$\frac{P(T|x)}{1 - P(T|x)} = e^{x\beta + \epsilon}, \tag{2}$$

while the probit model is defined by

$$P(T|x) = \frac{1}{\sqrt{2\pi}} \int_{-\infty}^{x\beta} e^{-z^2/2}\, dz. \tag{3}$$

Perl estimates the linear model by ordinary least squares and the logit and probit model by maximum likelihood techniques. In each case, individual households provide the unit of observation, so that the data set used in estimation contains 36,703 observations.

Three price variables are included in the models as predictors, the monthly service charge, the service-connection charge for a new subscriber, and a dummy variable that distinguishes whether or not a household resides in an area with measured local service.[9] The monthly service charge used by Perl is adjusted for the level of income in order to approximate the minimum charge at which a household can obtain telephone service.[10] The dummy variable

---

[7] Perl notes that another drawback of the linear model is that the least squares estimates of the coefficients in β are inefficient because the variance of the error term ϵ depends on the values taken by x. However, in this case, a generalized least squares estimator can be used that is asymptotically efficient. See Goldberger (1964, pp. 248-50).

[8] See McFadden (1974) for a discussion of conditional logit models, and their estimation by maximum likelihood methods. The standard reference on the probit model is Finney (1971).

[9] The price data are based on information from AT&T's Market Research Information System (MRIS), which contains billing records and household data for 30,000 residential subscribers located in each of one hundred Bell System Revenue Accounting Office (RAOs). (An RAO is a collection of exchanges, usually contiguous, that have been grouped together for accounting purposes.) The MRIS data on monthly service charges were linked to Census data by identifying the RAO in which each Census household was located.

[10] See Perl (1978, pp. 10-11) for details. The purpose of this adjustment is to neutralize the positive correlation between income and the monthly ser-

for the presence of measured service is intended to allow for the lower monthly service charge that is likely to result from local use (or at least a part thereof) being charged for directly.

Besides the three price variables, the other predictors in Perl's models include household income, age of household head, years of schooling of household head, size of household, and dummy variables indicating whether or not the head of the household is employed; the household resides in an urban area; the household resides in the South; the head of the household is black; the household is headed by a male, no spouse present; one-person household, male; one-person household, female.

Perl finds that the demand for basic telephone service is significantly influenced by both the monthly service charge and the service-connection charge. His estimates of the average elasticity of demand with respect to the monthly service charge range from −0.067 to −0.091 and from −0.016 to −0.022 with respect to the service-connection charge. Further, the price elasticity of demand is found to be quite sensitive to the level of household income and to the age of the head of the household.[11] The logit and probit models appear to perform somewhat better than the linear model. The average impact of specific variables is similar in the three models, but price and income effects are larger in the logit and probit models. The percentage of households with basic telephone service is found to be positively related to income, age, and education, and inversely related to the number of persons in the household. This percentage is also higher for households having its head employed, higher in urban than in rural area, lower in the South than elsewhere, and lower for nonwhite households. Finally, the percentage of house-

---

vice charge. This positive correlation reflects the fact that as income rises, families often choose a more enriched grade of basic service, such as one-party service as opposed to two- or four-party service. As a consequence of this positive correlation between income and the basic service charge, average charges are likely to appear higher in high income areas than in low income areas. Since high income areas are also likely to have higher telephone availability, this—as Perl noted in an earlier version of his paper—would tend to bias the relationship between telephone availability and telephone charges toward zero. Perl also analyzed two other measures of the basic monthly service charge, the average monthly service charge in the RAO in which the household was located and the average monthly charges paid by households with annual income under $5,000. However, of these three measures of price, Perl concludes (correctly, in my opinion) that the income-adjusted measure is the most appropriate, and bases most of his analysis on this measure.

[11]For example, for households with a twenty-year-old head and income of $1,500, Perl's estimates of the price elasticity with respect to the monthly service charge range from −0.173 to −0.422—three to four times the average elasticity for all age groups and income levels.

holds with basic telephone service is lower among single-person households (both male and female) and lower also for male-headed households with spouse absent than for other household types.

In general, the Perl study is an important contribution to the empirical analysis of access demand. It utilizes a rich data base—indeed, the study has no peers in this regard—and the findings are interesting and useful. However, as in the Feldman study, I find the analysis to be weakly motivated theoretically. Perl does an excellent job in identifying the charges corresponding to access and then relating the demand for basic service to them, but once again the dependence of the demand for access on the benefits conferred by use is ignored.[12] The impact of system size is also ignored. However, since Perl distinguishes between urban and rural households, and since urban exchanges are larger than rural exchanges, the system externality is reflected, at least to some extent, in the higher demand by urban households.

The final study to be reviewed in this section is by Pousette (1976). Pousette estimates access and use equations for residence, business, and public sector customers in Sweden using aggregate time-series data for the period 1949-1974. Pousette does not explain the number of subscribers directly, but rather specifies separate equations for new connections and disconnections.[13] The "stock" of telephones is then calculated as last period's stock plus new connections minus disconnections. Pousette's overall model therefore consists of behavioral equaitions for telephone use, new connections, and disconnections, plus the stock identity.

To illustrate Pousette's analysis, we shall present his equations for residence new connections and disconnections. The equation for new connections is as follows (t-ratios appear in parentheses):[14]

$$\begin{aligned}\hat{CT}_R = &\underset{(-0.22)}{-20.77} + \underset{(3.00)}{0.09}(H - ST_R) + \underset{(1.26)}{3.78}\left(\frac{C}{H}\right) - \underset{(-2.78)}{0.50}\left(\frac{P_U}{P_C}\right)\\ &- \underset{(-9.54)}{1.24}\left(\frac{P_I}{P_C}\right) + \underset{(7.85)}{1.57}AUT_L \qquad \overline{R}^2 = 0.970,\end{aligned} \tag{4}$$

[12] While it may not lead anywhere, it would be interesting to include a toll price index as a (negative) measure of the benefits generated by toll calls.

[13] In the telephone industry in the United States, reconnections and new connections and disconnections are referred to as inward and outward movement. Since I shall be using Pousette's notation in presenting his equations, it is also more convenient to keep his terminology.

[14] Because there was an installation backlog during the sample period in Sweden, new connections refer to telephones ordered rather than actual in-

where

$\hat{CT}_R$ = number of new residence connections ordered (in thousands)
H = number of households (in thousands)
ST_R = stock of telephones in residence (in thousands)
C = private consumption (in thousands of 1968 kronor)
P_U = weighted sum of subscription charge and call charge
P_C = implicit deflator for private consumption
P_I = service-connection charge index
AUT_L = index for the automation of local calls.

For residence disconnections, Pousette obtains:[15]

$$\hat{DT}_R = -165.53 + \underset{(4.58)}{3.50D} + \underset{(5.40)}{1.62E} + \underset{(3.25)}{0.26CT_R} \quad (-6.30)$$

$$+ \underset{(5.28)}{21.75D4960} + \underset{(2.90)}{14.45D61}$$

$$\bar{R}^2 = 0.924, \tag{5}$$

where

$\hat{DT}_R$ = number of disconnections ordered (in thousands)
D = number of deaths, not married, twenty years or over (in thousands)
E = number of emigrants, twenty years and over (in thousands)[16]
CT_R = number of residence new connections ordered (in thousands)
$D4960$ = dummy variable for incorrect data (1 for 1949–1960, 0 otherwise)
$D61$ = dummy variable for incorrect data (1 for 1961, 0 otherwise).[17]

---

stallations. Disconnections also refer to orders, although in this case queuing was not a problem.

[15] While Pousette uses both ordinary least squares and three-stage least squares in estimation, only the OLS results are reported here.

[16] The number of married emigrants is divided by 2.

[17] The two dummy variables correct for a change in definition in the time series for disconnections in 1961. The dummy variable for 1961 is, of course, tantamount to eliminating that year from the analysis (except that the zero residual for the year goes into the calculation of the $\bar{R}^2$).

Income is represented in the equation for new connections by private consumption, and is the only predictor with a *t*-ratio less than 2 (in absolute value). The term $H - ST_R$, which denotes the number of households without a telephone, is included as a measure of market potential. The service-connection charge is the most significant predictor, followed by the index for the automation of local calls. The highly significant positive coefficient on this variable attests that service-quality is an important factor and that Swedes place a positive value on their time. The price index for subscription and call charges on new connections is less than that of the service-connection charge. In terms of elasticities (which Pousette unfortunately does not provide), it would appear that the elasticity with respect to the subscription and use charges is about half the elasticity for the service-connection charge.

In the equation for disconnections, deaths and emigration are the most important predictors.[18] The number of new connections is included in the equation as a measure of the "potential" for disconnections. Pousette observes that the longer that a household has had a telephone, the more difficult it is likely to be to get along without the service. The best candidates, therefore, for disconnecting are those that have just recently become subscribers. The positive, and highly significant, coefficient on new connections supports this view.

Pousette's study is of a great deal of interest because it, along with Waverman (1974), comes closest to utilizing the access/use framework outlines in Section 3 of Chapter 2.[19] Pousette recognizes that the demand for access depends on the price of use, as well as the subscription fee and the service-connection charge, although it is unfortunate that the subscription fee and the price of use are combined in a single index. Pousette also departs from the model suggested in Chapter 2 in that the focus is on the charge in access demand, as opposed to the level of access demand. But by specifying separate equations for new connections and disconnections, Pousette is able to focus more explicitly on sociodemographic factors. Interestingly enough, though, this is probably of greater relevance in the U.S. market, where changes in address are frequent and there is substantial seasonal "churning."

---

[18] Swedes do not change residences with anything approaching the frequency of Americans. The coefficient for the number of deaths, it will be noted, is 3.50, which seems high.

[19] Waverman analyzes data for Canada and the U.K., as well as for Sweden. Waverman's analysis differs from Pousette's in that the stock of telephones

Pousette's specifications do have some problems, however, in that the dependent variables are changes, while the independent variables are levels. Of course, it may be that the relationship between the stock of telephones and price and income is quadratic, in which case Pousette's specifications are appropriate. I doubt, though, that this is what Pousette had in mind. Also, Pousette does not make any allowance for the access externality in his equation for new connections. The existing number of subscribers does appear in the market potential term $H - ST_R$, and one might interpret the negative coefficient on ST_R (-0.09) as reflecting a nonlinearity in the access externality. But this would seem to be pushing the point. Pousette interprets $H - ST_R$ as a measure of nonsaturation, and the empirical results have to be viewed as bearing this out.

In Table 1-1, I have listed the price and income elasticities of the demand for access that have been estimated in the empirical studies. The conclusion that emerges is that the price elasticity, calculated with respect to the price of basic service, appears to be in the neighborhood of -0.1, while the income elasticity appears to be in the neighborhood of 0.5. Waverman's results for residence main stations in Ontario and Quebec and Perl's results for the United States suggest that the income elasiticity may be smaller than 0.5, but these are the only exceptions.[20] A very small price elasticity and a moderate, but yet decidedly inelastic, income elasticity is precisely what one should expect for basic telephone service: Access to the telephone system is not a plaything of the rich (at least in the developed countries), but has become a basic necessity for virtually all income groups.

Perhaps the most important finding is that the price elasticity of the demand for access is in fact different from zero. The results of Perl are especially strong in this regard, and while one might question certain aspects of Perl's models, his price results will almost certainly hold up to respecification. As one should expect, the elasticity with respect to the service-connection charge is smaller than the elasticity with respect to the subscription fee. Perl's estimate of this elasticity is -0.02, while Waverman's estimate is -0.04. In neither case is the estimate strong statistically, which contrasts

---

is explained directly, and local calls are distinguished from intercity calls. Waverman's equation for local use will be discussed in the next section.

[20] In both cases, the estimate of the income elasticity is 0.15. Perl's elasticity (calculated from his logit equation) is estimated with a model in which other factors such as education, which is highly correlated with income, are taken into account. Thus a smaller income elasticity is probably to be expected.

Table 3-1. Estimates of Price and Income Elasticities of the Demand for Access to the Telephone System

Class of Customer and Study	*Dependent Variable*	*Price Elasticity*		*Income Elasticity*	*Type of Data*
		Service-Connection Charge	*Basic Service Charge*		
Residential					
Alleman	main stations	NE	−0.17	0.56	CS: Cities, U.S.
Feldman	main stations plus extensions	NE	−0.05	0.54	CS: States, U.S.
Perl	telephone availability	−0.02	−0.08	0.15	CS: Households, U.S.
Rash	main stations	NE	−0.11	0.61	TS: A, Ontario and Quebec
Waverman	main stations	NE	−0.12	0.15	TS: A, Ontario and Quebec
Business					
Waverman	main stations	NE	−0.09	NE	TS: A, Ontario and Quebec
Residential & Business Combined					
Davis et al.	total telephones less residence main stations	NE	−0.08	0.39	TS: A, Bell System, U.S.
Waverman	main stations	−0.04	−0.06	0.56	TS: A, Sweden

Symbols: NE: not estimated; CS: cross-section; TS: time series; A: annual.

sharply with the results of Pousette. Pousette found [cf. equation (4)] the service-connection charge to be considerably stronger statisticaly than the subscription fee. And while Pousette's actual elasticities are not available, it is clear from the regression coefficients that the elasticity for the connection charge is at least double the elasticity for the subscription fee.

For Perl and Waverman, however, the quantity being explained is either telephone availability or the number of subscribers, while for Pousette, the quantity being explained is the number of new subscribers. Because the service-connection charge is faced only once—at the time of connection—we should naturally expect this charge to have more of an impact on additions to the stock of telephones than on the stock itself.[21] Thus, that the service-connection charge shows up much stronger in Pousette's equation than in the equations of Perl and Waverman is to be expected. Moreover, that the elasticity for the service-connection charge is larger than the elasticity for the subscription fee suggests that, even in Sweden, there may be a population of "churners" whose turnover of main stations can be dampened by higher charges for reconnection.

In closing this review of access demand, I would like to reproduce, in Table 3-2, a part of a table from Perl's study that shows the dependence of the estimated elasticity with respect to the price of basic service on the level of income and the age of the head of household. Perl calculated these estimates for all three of his models, linear, logit, and probit, but, since the three models yield similar results, Table 3-2 contains only the estimates from the logit model.

Table 3-2 tells an interesting story. The estimated elasticity of the demand for basic service with respect to the monthly charge for that service varies directly with the size of the charge, inversely with the level of household income, and inversely with the age of household head.[22] The first two results are really a reflection of the same thing, namely, the importance of the income effect in the overall price elasticity for access demand. Let us concentrate for a moment on households whose heads are thirty-five years of age.

---

[21] It has been suggested by Paul Brandon of Bell Labs that the service-connection charge should be amortized over several years as part of the subscription fee. Ed Lowry of AT&T has taken a few tentative research steps in this direction as part of his Ph.D. dissertation, but the results so far are not very encouraging. Neither are the results of Robinson and Fask (1977).

[22] The interaction terms upon which the estimates of Table 3-2 are based are all highly significant statistically.

Table 3-2. Estimated Elasticity of Demand with Respect to the Income-Adjusted Basic Monthly Service Charge (Perl Study, Logit Model)

Age of Household Head and Household Income	Income-Adjusted Monthly Service Charge		
	$4.00	*$5.75*	*$7.50*
20 years of age			
$1,500	−0.168	−0.282	−0.422
$4,500	−0.129	−0.220	−0.337
$7,500	−0.093	−0.161	−0.251
$10,500	−0.064	−0.113	−0.178
$17,500	−0.026	−0.045	−0.074
35 years of age			
$1,500	−0.131	−0.223	−0.340
$4,500	−0.096	−0.165	−0.258
$7,500	−0.064	−0.112	−0.177
$10,500	−0.042	−0.074	−0.119
$17,500	−0.013	−0.023	−0.037
55 years of age			
$1,500	−0.089	−0.154	−0.242
$4,500	−0.062	−0.109	−0.171
$7,500	−0.040	−0.071	−0.114
$10,500	−0.026	−0.046	−0.074
$17,500	−0.010	−0.018	−0.029
70 years of age			
$1,500	−0.066	−0.116	−0.184
$4,500	−0.043	−0.075	−0.120
$7,500	−0.027	−0.049	−0.080
$10,500	−0.017	−0.031	−0.051
$17,500	−0.008	−0.014	−0.023

Source: Perl (1978, Table 7).

The value of −0.013—the estimated price elasticity for a monthly service charge of $4.00 for a household income of $17,500—should lie reasonably close to the pure substitution elasticity (i.e., the Slutsky, income-compensated price elasticity) for a household with this age of head. Compare this number to the value of −0.340 for the elasticity for a household with head of age thirty-five, income only $1,500, and facing a monthly service charge of $7.50. The income effects are relatively very large.

When substitution effects are large relative to income effects, consumers can substitute away from goods whose prices have risen with little loss in utility. However, when income effects are large relative to substitution effects, an increase in price means a relatively large decrease in utility. Since the income effect is indicated to be large relative to the substitution effect in the price elasticity of demand for access for households with low income, particularly if they are young, the welfare of these households may be significantly decreased by increases in the price for basic service. The results of Perl thus suggest that the telephone companies may encounter spirited customer opposition to moves to bring the price of access for residential customers into line with the cost of supply.

THE DEMAND FOR LOCAL USE

We shall now turn our attention to the demand for local use. Unfortunately, the demand for local use is difficult to analyze empirically because the price of local use is usually rolled into the monthly service charge. As a consequence, much of the empirical literature on local demand involves the analysis of local revenues, which is not coextensive with local use. However, there are a few studies—Beauvais (1977), New York Telephone Co. (1976a, 1976b), Pavarini (1975, 1978), and Waverman (1974)—that have been able to analyze the demand for local use directly. Most of this section will be devoted to discussion of these studies.

We shall begin with the Beauvais study, which provides an interesting contrast of innovations and problems. Beauvais is concerned with estimating the demand for outgoing local calls by residential customers under flat-rate pricing, with the view of being able to estimate the impact on residential customers of a move to measured service from flat-rate service. Beauvais's analysis is based on data from GT&E's measured service trial in three exchanges in Illinois. The data represent a one-month (August 1976) cross-section of observations from a sample of about 1,000 residential subscribers in these three exchanges.

Beauvais assumes that the number of outgoing local calls is a function of income, the marginal price of a local call, and the average price of a local call. The actual level of a household's income was not collected, but only the interval within which the household's income fell. Income is accordingly represented by a set of dummy variables. The average price of a local call is measured by the monthly service charge divided by the number of outgoing local calls. Unfortunately, defining the average price in this way involves a serious error, and essentially vitiates Beauvais's empirical results. But, more about this below.

The marginal price of a local call is measured by the opportunity cost of the time spent by a caller in making a call. Beauvais measures this opportunity cost as the amount of time spent on a call (by all subscribers in the sample) multiplied by the subscriber's wage rate. While one may question whether this is the appropriate way of defining the opportunity cost of a call, the recognition that the price of a call involves more than just the out-of-pocket expense of the call (which, in this case, is zero) is a major innovation. I am not aware of any study in the literature that has preceded Beauvais in doing this.

But, while the introduction of the opportunity cost of time represents a major innovation, defining the average price as the monthly service charge divided by the number of outgoing calls creates a major problem. As Beauvais himself notes, in having the dependent variable (i.e., the number of out-going local calls) in the denominator of the average price, a negative relationship between the number of calls and the average price is necessarily established, which is to say that the estimate of the coefficient on the average price will be biased downward. The strength of this bias will depend, in great part, upon the variance of the monthly service charge in relation to the variance of the number of calls. If the latter is large relative to the former (as is almost certainly the case), the bias *can* be substantial. And, if the number of calls is only weakly correlated with the monthly service charge, the bias *will* be substantial.

In the data analyzed by Beauvais, it is unlikely that there is much variation in the monthly service charge, but it is very likely that there are a few households that make a very large number of calls.[23] The presence of just a few outliers of this type is sufficient to impart a substantial negative bias to the estimate of the coefficient on the average price.[24] What Beauvais might have done was to include the monthly service charge as a predictor without dividing it by the number of outgoing calls. This would yield a model that would be in keeping with the treatment of two-part tariffs discussed in Section 5 of Chapter 2, but with the added feature that the opportunity cost of a caller's time is explicitly taken into account.

Let me now turn to Waverman's study. Waverman analyzes the demand for local use in Sweden in a framework that draws on the access/use distinction discussed earlier. The model postulated for explaining local usage is

$$ln \left(\frac{Q}{T}\right)_t = b_0 + b_1 \; ln \left(\frac{Q}{T}\right)_{t-1} + b_2 ln X_t + b_3 ln P_t + u_t, \qquad (6)$$

---

[23] Cf. Garfinkel, L. (1977a), (1977b), and (forthcoming).

[24] I was alerted to the possibility of this problem in Beauvais's analysis by the size of the *t*-ratio for the coefficient on the average price. Its value is 67.6, while the next largest *t*-ratio is 10.9.

where

Q = number of local "pulses"[25]
T = number of telephones (main stations)
X = real gross domestic product/household
P = marginal price per pulse
u = random error term.

As an alternative to the model in (6), Waverman also estimates

$$\ln Q_t = b_0 + b_1 \ln Q_{t-1} + b_2 \ln X_t + b_3 \ln P_t + b_4 \ln T_t + u_t. \tag{7}$$

Waverman estimated his models using annual time-series data for the period 1949–1969. The results obtained for the two local-usage equations are as follows (t-ratios are in parentheses):

$$\ln \left(\frac{Q}{T}\right)_t = \underset{(3.50)}{1.15} + \underset{(1.43)}{0.294} \ln \left(\frac{Q}{T}\right)_{t-1} + \underset{(3.32)}{0.832} \ln X_t - \underset{(-1.50)}{0.097} \ln P_t$$

$$\bar{R}^2 = 0.971 \tag{8}$$

$$\ln Q_t = \underset{(-1.32)}{-4.17} + \underset{(1.45)}{0.286} \ln Q_{t-1} + \underset{(0.54)}{0.228} \ln X_t - \underset{(-2.33)}{0.272} \ln P_t + \underset{(3.36)}{1.19} \ln T_t$$

$$\bar{R}^2 = 0.997. \tag{9}$$

To begin with, it will be noted that the coefficient on $\ln T_t$ in equation (9) is greater than 1 (although not significantly so). This result suggests that, *ceteris paribus*, more calls are made per telephone in a large system, and can therefore be interpreted as a reflection of the access externality. Secondly, it is somewhat surprising to find that, with the stock of telephones held constant, habit formation (or inertia) in local calling is rather modest. That this is so is reflected in coefficients on the lagged dependent variable in equations (8) and (9) that are closer to 0 than 1, and, indeed, by conventional criteria, not significantly different from 0. Finally, it is seen that GDP/household is statistically significant in equation

---

[25] Many European telephone companies use a pulse system for measuring holding time, so that the number of pulses for a call represents the holding time (or length) of the call. For a discussion, see Mitchell (1978b).

(8), but not in equation (9). This no doubt reflects a collinearity between GDP/household and the number of main stations.

The final study of local-service demand to be given detailed review is Pavarini (1975). Pavarini is concerned with analyzing the impact on local calling of a conversion from flat-rate to measured service. The focus is on data taken from customers subscribing to METROPAC, an optional extended area service offered in some exchanges bordering the Denver metropolitan area. METROPAC was converted from flat rate to measured rate in January 1971. Pavarini's data set consists of observations on a sample of 387 residential customers and eighty business customers for three different four-month periods: April-July 1970, August-November 1970, and April-July 1971. The purpose of the study is to quantify the impact of the METROPAC conversion on local usage.

Ordinarily, one uses a standard demand model to estimate price and income effects, but in this case, the variation in price is too limited for a conventional demand model to be of much value. Pavarini uses, instead, a model that relates the number of calls after the conversion to measured service to the flat-rate demand and the measured-service price per call. The flat-rate demand is viewed as providing the base from which the impact of measured service is estimated. The standard formulation of this model would be to regress the number of calls after the conversion on the number of calls before the conversion and the price per call after the conversion. However, the conventional regression model is not really appropriate in this situation because of the fact that the flat-rate demand is itself subject to random variation. The independent variable is thus measured with error, and, as is well known, ordinary least squares applied to such a model will yield biased and inconsistent estimators. An errors-in-variables model is called for.

Pavarini does in fact adopt an errors-in-variables framework, although it is not of the form traditionally employed in econometrics, and Pavarini does not even refer to it as an errors-in-variable model. Briefly stated, Pavarini's methodology consists of estimating the "normal" variation in usage between the period August-November and April-July 1971 that would be expected to occur in the absence of the rate conversion. The difference between the estimated normal usage and that which was actually observed is then attributed to the rate conversion.[26]

---

[26] Pavarini's approach is similar in spririt to one developed in a recent article by Blomqvist (1977).

We shall begin with a description of Pavarini's method for estimating variation in normal usage. Suppose that there are two periods, and assume that a household's usage in each of the periods is a random variable. Let X and Y denote usage in the two periods, and let $f_{XY}(x,y)$ denote their joint distribution. Let $f_{Y|X}(y|x)$ denote the conditional distribution of y given x.[27] Finally, let $f_X(x)$ and $f_Y(y)$ denote the marginal distributions. By definition, $f_Y(y)$ will be given by

$$f_Y(y) = \int_x f_{Y|X}(y|x)\, f_X(x)dx. \tag{10}$$

In the present context, however, our interest is confined to the conditional distribution $f_{Y|X}(y|x)$.

Let us assume that we observe a particular value of usage in period 1, say $X = x^*$. What can we say about the value of y, which is yet to be observed? Clearly, the natural quantity to consider is the expected value of Y conditional on $X = x^*$. If the form of the conditional distribution $f_{Y|X}(y|x)$ is known, this conditonal mean can be calculated as

$$E(Y|x^*) = \int_y y f_{Y|X}(y|x^*)dy. \tag{11}$$

Now, let

U_{f1} = usage (measured as the number of minutes/month) in the period August-November 1970
U_{f2} = usage in the period April-July 1970
U_m = usage in the period August-November 1971.

In line with expression (11), Pavarini seeks to estimate the expected value of U_m given u_{f1} (i.e., given the observed amount of usage in the period August-November 1970) and given the conversion to measured service in January 1971. However, in order to allow for normal customer volatility, the period April-July 1970 is used as an intermediate step in the estimation.

Pavarini postulates that the conditional density of U_{f2} given u_{f1} [$f_{Y|X}(y|x)$ in the previous notation] will be lognormal with mean $\alpha(u_{f1})$ and variance σ^2.[28] Further, it is assumed that the

---

[27] It is assumed that $f_{Y|X}(y|x)$ is the same for all households.

[28] $(U_{f2}|u_{f1})$ being lognormal means that $1n(U_{f2}|u_{f1})$ is normal with mean $\alpha(u_{f1})$ and variance σ^2. The standard reference on the lognormal distribution is Aitchison and Brown (1957). Pavarini cites Johnson and Kotz (1970).

mean of $(U_{f2}|u_{f1})$ will be given by

$$E(U_{f2}|u_{f1}) = su_{f1}^{q} \tag{12}$$

and that

$$Var\ (U_{f2}|u_{f1}) = w^2 (su_{f1}^{q})^2, \tag{13}$$

where s, q, and w^2 are constants. It then follows from the properties of the lognormal distribution that the mean and variance of $ln(U_{f2}|u_{f1})$,$\alpha(u_{f1})$ and σ^2, will be equal to

$$\alpha(u_{f1}) = lns + q\, lnu_{f1} - \frac{ln(w^2 + 1)}{2} \tag{14}$$

$$\sigma^2 = ln(w^2 + 1). \tag{15}$$

It will be useful to rewrite expression (14) as

$$\alpha(u_{f1}) = \theta_1 + \theta_2\, lnu_{f1}, \tag{16}$$

where

$$\theta_1 = lns - \frac{ln(w^2 + 1)}{2} \tag{17}$$

and

$$\theta_2 = q. \tag{18}$$

Since $ln(U_{f2}|u_{f1})$ is assumed to be normally distributed with mean equal to $\theta_1 + \theta_2\, lnu_{f1}$ and variance equal to σ^2, estimates of θ_1, θ_2, and σ^2 can be obtained from a least-squares regression of lnu_{f2} on lnu_{f1}. From these estimates, it is possible to derive estimates of the parameters w^2, s, and q,[29] and one can then obtain an estimate of the expected value of $(U_{f2}|u_{f1})$ from

---

[29] Since $\theta_2 = q$, $\hat{\theta}_2$ will provide an unbiased estimator of q. However, since s and w^2 are nonlinear functions of θ_1 and θ_2, the estimators of s and w^2 obtained directly from $\hat{\theta}_1$ and $\hat{\theta}_2$ will, in general, be biased. Pavarini obtains asymptotically unbiased estimators of s and w^2 by utilizing the asymptotic distributions of $ln\hat{s}$ and $ln(\hat{w}^2 + 1)$.

$$\hat{E}(U_{f2}|u_{f1}) = \hat{s}u_{f1}^{\hat{q}}. \tag{19}$$

Pavarini notes that the foregoing, which provides a methodology for forecasting variation in "normal" usage under a flat-rate tariff, feeds into the larger problem of estimating the impact of conversion to measured service as follows:

> Instead of interpreting the price response as the transformation of U_{f1} to U_m, let us isolate the pure price response by considering that each customer gets from u_{f1} *to* u_m in two discrete steps. First, his usage changes to *what it would have been* had there been no price change. This is normal usage variation. Then, consider that it is this usage level that he adjusts so as to arrive at u_m. This second movement is the price effect. In the absence of any strong evidence of long-term growth in METROPAC usage, we will use U_{f2} (usage in the same period one year before) as a surrogate for the results of movement from U_{f1} due to normal usage variation. That is, U_{f2} will be the prediction of what usage would have been in the April-July 1971 period had there not been a price change. This is *not* to say that we will simply compare the values of i_{f2} and u_m for each customer, for such a comparison would still be "noise" contaminated. Rather, for customers who showed a specific value of usage just previous to the rate change (u_{f1}), a prediction is made of the probability density of their usage ($U_{f2}|$ u_{f1}) in the absence of price change and this density is transformed to the conditional density of ($U_m|u_{f1}$) via a price relation that has u_{f2} as its usage argument. [Pavarini (1975, pp. 16--17), italics in original.][30]

We now bring the conversion to measured service into the picture. Let $g(x,p)$ denote the number of calls made by customers with flat-rate demand x after they are charged a price of p per unit of usage. The transformation from U_{f1} to U_m, under Pavarini's assumptions, will therefore be given by

$$(U_m|u_{f1}) = g[(U_{f2}|u_{f1}),p]. \tag{20}$$

---

[30] As Pavarini observes (p. 17), the validity of this analysis requires that the conditional distribution describing the movement in usage from the period August-November 1970 to April-July 1970 also describes the movement (in the absence of a rate change) from August-November 1970 to April-July 1971. This requires (1) that there be no year-to-year trend in usage, (2) that any seasonal component be stable and represented in the parameters θ_1 and θ_2, and (3) that σ^2 be constant through time.

Pavarini considers two representations of the function g that preserves the lognormality of $(U_m | u_{f1})$,[31] one deterministic and the other stochastic. For the deterministic representation, it is assumed that

$$u_m = c(p) u_{f2}^{\,d(p)}, \tag{21}$$

while, for the stochastic representation, it is assumed that

$$(U_m | u_{f2}) = e^{A(p)} u_{f2}^{\,b(p)}, \tag{22}$$

where $A(p)$ is a random variable with mean $\mu_a(p)$ and variance $\sigma_a^2(p)$.

In equation (21), the functional forms of $c(p)$ and $d(p)$ cannot be specified because their values can only be estimated at $p = 0$, which corresponds to $c(o) = 1$ and $d(o) = 1$, and at the METROPAC price. However, since

$$\begin{aligned}(U_m | u_{f1}) &= g[(U_{f2} | u_{f1}), p] \\ &= c(p)(U_{f2} | u_{f1})^{d(p)},\end{aligned} \tag{23}$$

and because $(U_{f2} | u_{f1})$ is lognormal with mean $\alpha(u_{f1})$ and variance σ^2, it follows that $(U_m | u_{f1})$ will be lognormal with mean $d(p)\alpha(u_{f1}) + \ln c(p)$ and variance $[d(p)]^2 \sigma^2$.

Since we have estimates of $\alpha(u_{f1})$, estimates of $c(p)$ and $d(p)$ can be obtained from the coefficients in a least-square regression of $\ln u_m$ on $\hat{\alpha}(u_{f1})$,

$$\ln u_m = \eta_1 + \eta_2 \hat{\alpha}(u_{f1}) + \epsilon, \tag{24}$$

where

$$\eta_1 = \ln c(p), \tag{25}$$

$$\eta_2 = d(p), \tag{26}$$

[31] Pavarini establishes (by analysis of the residuals in a least-square regression of $\ln u_m$ on $\ln u_{f1}$) that the conditional density of $(U_m | u_{f1})$ can in fact be considered to be lognormal.

and ϵ is a random error term. Estimation of equation (25) will also yield an estimate of the variance of lnu_m, say $\hat{\gamma}^2$. However, from equation (21), it follows that

$$\gamma^2 = [d(p)]^2 \sigma^2. \tag{27}$$

Consequently, since we already have an estimate of σ^2, we have three equations [(25), (26), (27)] in the two unknowns $c(p)$ and $d(p)$. This provides a check on the consistency of the deterministic specification in (21).

In both the residence and business samples, Pavarini finds $\hat{\eta}_2 = \hat{d}(p) < 1$ and $\hat{\gamma}^2 > \hat{\sigma}^2$. Pavarini notes (p. 21): "Aside from the obvious inconsistency, this is a clear indication that the variance of the conditional densities is greater than one would expect if the price relation [were] deterministic. The results indicate that part of the observed variance is due to the different price-related reactions of customers at the same flat-rate usage level." Pavarini accordingly turns his attention to the stochastic relation in expression (22).

In (22) Pavarini assumes that, for fixed p, $A(p)$ is normally distributed with mean $\mu_a(p)$ and variance $\sigma_a^2(p)$. $A(p)$ is assumed to be independent of U_{f1}, and $b(p)$ is a constant. For $p = 0$, both $\mu_a(o)$ and $\sigma_a^2(o)$ are assumed to be 0, and $b(o)$ is assumed to be 1. Since[32]

$$e^{A(p)} \sim \Lambda[\mu_a(p), \sigma_a^2(p)] \tag{28}$$

and

$$(U_{f2} | u_{f1})^{b(p)} \sim \Lambda[b(p)\alpha(u_{f1}), [b(p)^2]\sigma^2], \tag{29}$$

it follows that

$$(U_m | u_{f1}) \sim \Lambda[\mu_a(p) + b(p)\alpha(u_{f1}), \sigma_a^2(p) + [b(p)]^2 \sigma^2]. \tag{30}$$

The unknown parameters in this expression, $\mu_a(p)$, $b(p)$, and $\sigma_a^2(p)$, can be estimated, as before, from a least-squares regression [cf. equation (24)] of lnu_m on $\hat{\alpha}(u_{f1})$. The estimates obtained by

[32] The notation $x \sim \Lambda(\mu, \sigma^2)$ means that x is distributed lognormally with mean μ and variance σ^2.

Pavarini, together with estimated standard errors, at the METROPAC price are given in Table 3-3.[33]

At this point, Pavarini is able to estimate the portion of the observed variation in usage between August-November 1970 and April-July 1971 induced by the conversion to measured service in January 1971. The price-stimulated portion, considered as the movement from u_{f2} to u_m, is found from the relation

$$(U_m \mid u_{f2}) = e^{A(p)} u_{f2}{}^{b(p)}, \tag{31}$$

using estimated values of μ_a, σ_a^2, and b. In particular, since $(U_m \mid u_{f2})$ is distributed lognormally, we will have

$$(U_m \mid u_{f2}) \sim \Lambda[\mu_a(p) + b(p) \ln u_{f2}, \sigma_a^2(p)]. \tag{32}$$

The expected value of $(U_m \mid u_{f2})$ as a function of flat-rate usage will then be equal to

$$E(U_m \mid u_{f2}) = e^{[\mu_a(p) + \sigma_a^2(p)/2]} \mu_{f2}{}^{b(p)}, \tag{33}$$

while the variance of $(U_m \mid u_{f2})$ will be equal to

$$\mathrm{Var}(U_m \mid u_{f2}) = [E(U_m \mid u_{f2})]^2 [e^{\sigma_a^2(p)} - 1]. \tag{34}$$

A point estimate of the change in usage induced by the rate conversion can be obtained from expression (33) using the estimated values of μ_a, σ_a^2, and b from Table 3-3.

The variation is usage that is actually observed (U_m versus U_{f1}) is specified by "adding" the normal variation in usage U_{f2} versus

Table 3-3. Estimates of the Parameters in Equation (30)

	$\hat{b}(p)$	$\hat{\mu}_a(p)$	$\hat{\sigma}_a^2(p)$
Residence	0.41 (0.06)	2.08 (0.30)	0.28 (0.02)
Business	0.80 (0.10)	0.87 (0.60)	0.06 (0.04)

Source: Pavarini (1975, p. 23).

[33] In interpreting his results, Pavarini assumes that $\alpha(u_{f1})$ is measured without error.

U_{f1}) to the price-induced variation (U_m versus U_{f2}). Since the price relation specifies u_m as a function of u_{f2}, an estimate of u_{f2} is required as an input to the price model. Alternatively, to predict the distribution of usage following a rate change, both a price model (U_m versus U_{f2}) and knowledge of the distribution of U_{f2} are required.

Pavarini illustrates the implications of his errors-in-variables framework by comparing the predicted impact on the number of "measured" calls using his model, which allows for normal variation in the flat-rate demand, with the predicted impact from a model which does not allow for such variation. Briefly summarized, his results indicate that

> . . . the existence of normal usage variation causes the mean *total* response to be higher (lower) than the pure *price* response at low (high) flat rate levels. The price effect shows, for example, that customers with a flat rate demand of 200 minutes per month exhibit a mean measured usage of 82.6. However, had one interpreted the total effect as purely price-related, the estimate of the mean price response would have been some 7% higher (88.3). In this case, the 7% difference is "small" considering that movement was from the 200 level. However, as usage reduction due to price becomes less than the 59% reduction here (200 to 82.6), the contamination due to normal usage variation becomes more significant. In the limiting case when usage reduction due to price is *zero*, the total effect model would show a usage *increase* from the 200 level to a new mean level of 248. [Pavarini (1975, p. 26), italics in original.]

Measured local service is increasingly discussed these days, and Pavarini's model provides a vehicle for predicting what its impact on usage might be. It is for this reason that I have described his methodology in detail.[34] Also, one can also see Pavarini's methodology being used in analyzing the data being generated in the time-of-day electricity pricing experiments sponsored by the Federal Energy Administration (now the Department of Energy).[35] I mention the possibility of this because price variation is at a pre-

---

[34] In his 1976 paper, Pavarini estimates the impact on local usage of the METROPAC rate conversion for each of the three Denver exchanges separately (Erie, Ft. Lupton, and Longmont). Differences in estimated parameters across exchanges are then related to differences in income. In attempting this, Pavarini embeds his model in a utility-maximization framework. Readers interested in Pavarini's methodology should read *both* Pavarini (1975) and Pavarini (1976b). See also Pavarini (1979).

[35] For a description of these experiments, see Hill et al. (1979).

mium in most of these experiments, and Pavarini's methodology provides a useful alternative to the conventional regression model.[36]

So far, our focus in this section has been on the few studies that have analyzed the demand for local use per se. A number of other studies have looked at local service, but they have taken price-deflated local revenues as the dependent variable. What is being analyzed in these studies is a category of demand that combines access with local use. The models estimated may be useful for explaining real expenditures for local service, but since the distinction is blurred between access and use, little or no information is conveyed about the demand for local use.[37]

The studies that estimate equations for price-deflated local revenues include Davis et al. (1973), Dobell et al. (1972), New York Telephone Co. (1976a, 1976b), and Southern New England Telephone Co. (1977a). Davis et al., as part of the FORECYT model, have an equation for local service for residence and business combined that is estimated with aggregate quarterly time-series data for the Bell System. Dobell et al. employ annual time-series data, but the geographical entity is Canada. Equations for local-service revenues are estimated for residence and business customers combined using data for all of Canada, and residence customers and business customers separately using Bell Canada data for Ontario and Quebec. The SNET study uses quarterly time-series data for Connecticut and estimates separate equations for residence and business. Finally, the NYT study estimates equations for residence and business, both separately and combined, using quarterly time-series data for New York.

In Table 3-4, I have tabulated the estimated price and income elasticities of demand from the studies that have been discussed in this section. However, of the studies that focus explicitly on local use, only the estimates obtained by Waverman are included in the table. Pavarini's model does not readily allow for the calcu-

---

[36] Before moving on to the studies that analyze the demand for basic service, I should mention briefly the two works done by the New York Telephone Co. In the areas served by NYT, a specified number of local message units are included in the monthly service charge, but message units beyond the allowance cost the subscriber so much per message unit. The NYT studies analyze the demand for additional message units for residential customers, business customers, and for residential and business customers combined. The price and income elasticities estimated in these studies are tabulated in Table 3-4.

[37] Since in most exchanges local calls are free (not counting the opportunity cost of a caller's time), the demand for local service should more properly be interpreted as primarily a demand for access.

Table 3-4. Estimates of Price and Income Elasticities of the Demand for Local Use and Local Service

Type of Demand		*Price Elasticity*		*Income Elasticity*		
Residential	*Service*	*SR*	*LR*	*SR*	*LR*	*Type of Data*
Dobell et al.	basic	NE	NE	0.42	2.38	TS: A, Ontario and Quebec
NYT (1976a)	basic + terminal equipment	−0.10	−0.14	0.10	0.14	TS: Q, New York
NYT (1976a)	AMU	−0.35	NE	0.25	NE	TS: Q, New York
SNET (1977)[a]	basic	−0.02	−0.21	NC	0.16	TS: Q, Connecticut
Residential-Business						
Davis et al.	basic	−0.21	−0.27	0.25	0.33	TS: A, Bell System
Dobell et al.	basic	−0.20	−0.70	0.38	2.16	TS: A, Canada
NYT (1976b)	basic	−0.03	−0.17	0.07	0.40	TS: Q, New York
NYT (1976b)	AMU	−0.17	NE	0.29	0.67	TS: Q, New York
Waverman	basic	−0.27	−0.38	0.23	1.25	TS: A, Sweden
Business						
Dobell et al.	basic	NE	NE	0.99	1.55	TS: A, Ontario and Quebec
SNET (1977)[a]	basic	−0.03	−0.07	NE	NE	TS: Q, Connecticut
NYT (1976a)	basic + terminal equipment	−0.08	−0.15	0.10	0.19	TS: Q, New York
NYT (1976a)	AMU	−0.18	NE	0.13	0.24	TS: Q, New York

[a]The long run in the SNET study refers to the elasticities calculated after the lapse of one year.

Symbols: TS: time series; AMU: additional message units; CS: cross section; NE: not estimated; A: annual; NC: not calculated; Q: quarterly.

lation of a price elasticity, and I have omitted Beauvais' results because of the problem, noted earlier, with his definition of the average price of a local call. Dynamic models, usually of the Houthakker-Taylor or Koyck distributed-lag variety, are employed in all of the studies, so that both short-run and long-run elasticities have been estimated.

The estimated price elasticities are typically small, which is to be expected. The price elasticity for local use should be expected to be larger than the price elasticity for access, but it should not be too much larger because of the importance attached to the making and receiving of local calls. There is only one estimated price elasticity in the table that seems seriously open to question, namely, the Dobell et al. long-run price elasticity using data for all of Canada. The estimated value of −0.7 seems implausibly high, and is also completely out of line with the value of 0 found in their residence and business equations for Ontario and Quebec.

Apart from the Dobell et al. equation, the estimated elasticities in the equations for local service (in which price-deflated local revenue is the dependent variable) are generally in line with the price elasticities for access that are tabulated in Table 3-1. This provides support for the earlier observation that the demand for local service is primarily to be identified with the demand for access. The most interesting entry in the table, in my opinion, is Waverman's study, since it is the only one that focuses explicitly on the demand for local use. The price elasticity is estimated to be −0.27 in the short run and −0.38 in the long run. While I do not know at this point whether these elasticities are too large, too small, or just right in absolute value, they are larger than the price elasticities for access, and this provides some reassurance.

The estimates of income elasticity in Table 3-4 are much more varied than the estimates of price elasticity. Most of the estimated long-run income elasticities are less than 0.5, but the three equations of Dobell et al. all have long-run values in excess of 1.5. Once again, I feel that Waverman's results are to be given the most weight because of the explicit focus on local use. Waverman's estimate of the long-run income elasticity is 1.25, so that the estimates of Dobell et al. may be closer to the mark than those of the other studies.[38]

---

[38] Since Waverman's equation for local use is embedded in an access/use framework, the long run thus allows for a change in income to affect the number of telephones. The estimated income elasticity in Waverman's equation for the number of telephones is 0.56. The value of 1.25 is then obtained [from the coefficients in equation (9)] as $[0.228 + 1.19\ (0.56)/(1 - 0.286)]$.

TOLL DEMAND

We now turn to a discussion of toll demand, which constitutes by far the largest component of the empirical literature on telephone demand. The format will be to discuss interstate and intrastate toll separately. In classifying the literature on toll demand, I have defined interstate toll to include all studies of interstate toll in the United States and all studies of toll demand within national boudaries in other countries. Although the distinction between interstate and intrastate toll is artificial from the point of view of demand,[39] it has been retained because it assists materially in organizing the literature.[40] Only the most common form of toll service, namely, message toll service (or MTS), will be discussed in this section.

Interstate (Long-Haul) Toll Demand

The literature that we shall review under interstate toll includes AT&T Long Lines [Betteridge (1973)], Davis et al. (2973), Deschamps (1974), Dobell et al. (1972), Khadem (1973a), Kwok, Lee, and Pearce (1975), Larsen and McCleary (1970), Pousette (1976), Rash (1972), Waverman (1974), and Wert (1976). As with local-service demand, our procedure will be to describe briefly the main features of these studies and to tabulate the estimate of the price and income elasticities of demand. Five studies will be reviewed in detail.

A wide range of models and data are represented in the interstate toll studies. The extreme of both size and disaggregation is provided by the pioneering Long Distance Interstate (LDI) model of AT&T Long Lines, which disaggregates toll volume to 6750 different cells and offers the facility for predicting the impact on each of a change in the interstate toll schedule. Indeed, in comparison with the LDI model, the rest of the models in the literature are highly aggregated. The level of aggregation employed in the other studies is as follows: Davis et al., Deschamps, and Pousette all focus entirely on total toll volume. Dobell et al., Larsen and McCleary, and Waverman estimate models for residence and business customers separately, as well as for total toll. Finally, Khadem,

---

[39] What matters is price and length-of-haul, not that a state boundary is crossed.

[40] Most of the literature on toll demand relates to intrastate toll, which, in great part, reflects the widespread use in recent years of intrastate toll demand models in rate filings before state regulatory commissions.

Rash, and Wert restrict their attention to residential customers. Larsen and McCleary use cross-sectional data in estimation exclusively, while Deschamps and Wert use both time-series and cross-sectional data. The remaining studies all rely entirely on time-series data.

In general, the models that have been used for toll demand are similar to those that have been used for local service, and two broad classes of models can be identified. The first class approaches the dependence of toll volume on the stock of telephones in a distributed-lag framework. With these models, the stock of telephones appears in the analysis only implicitly. The models of Dobell et al., Khadem, and Rash fall into this category. With the second class of models, the stock of telephones appears explicitly, either as a deflator of toll volume or directly as an independent variable. The models of Davis et al., Pousette, and Waverman are of this type.[41]

Most of the focus in the interstate studies is on estimating price and income elasticities of demand for rather broadly defined customer classes. Nevertheless, some interesting twists are to be found. Wert, for example, is concerned with estimating the impact of day/night differentials in price on the day/night distribution of toll traffic. His conclusion is that, on the whole, given the time span investigated, changes in day/night price differentials lead to only minor shifts in traffic. Larsen and McCleary, on the other hand, are concerned with modeling interstate toll demand on a point-to-point basis. The dependent variable in their analysis is the directional traffic between state pairs. Finally, Deschamps focuses on toll demand in Belgium, using traffic volume by mileage bands as the dependent variable. As we shall see below, Deschamps' study is especially interesting in the way that price is analyzed.

The estimated price and income elasticities of demand obtained in the interstate studies are tabulated in Table 3–5. The elasticities listed for the LDI model are calculated with reference to a typical long-distance interstate call. As mentioned earlier, Pousette does not present sufficient information to calculate the elasticities implied in his toll equation. The elasticities appearing in Table 3–5 are, on the whole, fairly substantial, and this is true for price as well as for income. Toll price is missing from only one model in the table (the business equation of Dobell et al.), while income is

[41] The models of Davis et al. and Waverman employ a distributed-lag framework as well. Allowance is thereby made for inertia in adjusting the volume of toll calling, for a given stock of telephones, to a change in price or income.

Table 3-5. Estimates of Price and Income Elasticities of Demand for Long-Haul Toll Calls for the United States and Other Countries

Study and Type of Demand	Dependent Variable	Price Elasticity Short-Run	Price Elasticity Long-Run	Income Elasticity Short-Run	Income Elasticity Long-Run	Type of Data
Residential						
Dobell et al. (1972)	PDR	−0.30	−1.90	0.20	1.27	TS: A, Ontario and Quebec
Khadem (1973)	PDR	−1.28	−2.58	0.40	0.84	TS: Q, Trans-Canada
Larsen and McCleary (1970)	messages	NE	−1.01	NE		CS: M, States
Rash (1972)	PDR	[a]	−0.94	[a]	1.53	TS: M, Ontario and Quebec
Waverman (1974)	PDR/T	NE	−1.16	NE	1.03	TS: A, Ontario and Quebec
Wert (1976)	daytime calls (%)	(−0.29)		(1.06)		TS-CS: M, Bell System
Business						
Dobell et al. (1972)	PDR	NE		0.19	2.76	TS: A, Ontario and Quebec
Larsen and McCleary (1970)	messages	NE	−0.98	NE		CS: M, States
Waverman (1974)	PDR/T	−1.20	−1.35	0.45	0.51	TS: A, Ontario and Quebec
Residential-Business						
AT&T (LDI)	messages	NE	−0.4 to −0.5	NE	≈ 1.00	TS: Bell System
Davis et al. (1973)	PDR	−0.88	−1.03	0.83	0.96	TS: Q, Bell System
Deschamps (1974)	messages	(−0.24)		NE		TS: M, Belgium
Dobell et al. (1972)	PDR	−0.11	−2.57	0.08	1.90	TS: A, Canada
Kwok, Lee and Pearce (1975)	messages	−0.18	−0.41	0.51	1.15	TS: Q, Ontario and Quebec
Kwok, Lee and Pearce (1975)	messages	−1.70	−2.71	0.83	1.32	TS: Q, Trans-Canada
Waverman (1974)	PDR/T	−0.72	−1.12	1.00	1.56	TS: A, Great Britain
Waverman (1974)	messages/T	−0.41	−0.72	0.56	0.99	TS: A, Great Britain
Waverman (1974)	pulses/T	−0.51	−1.08	0.23	0.50	TS: A, Sweden
Waverman (1974)	messages/T	−0.29	−0.58	0.19	0.38	TS: A, Sweden

[a] Cannot be calculated.

Symbols: PDR: price-deflated revenue; M: monthly; T: number of main-station telephones; TS: time series; A: annual; CS: cross section; Q: quarterly; NE: not estimated.

omitted from two (Deschamps and Larsen-McCleary). In general, the evidence otherwise suggests long-run elasticities of at least 1 for income and in the neighborhood of −1 for price. The estimated price elasticities are substantially smaller than 1 (in absolute value) in the LDI model and in the residential-business equations of Deschamps for Belgium and Kwok et al. for Ontario and Quebec. However, Deschamps' model is static rather than dynamic—a dynamic model would almost certainly yield a larger long-run price elasticity—and Kwok et al. find much larger price elasticities using TransCanada data.[42] In view of the extensive and generally successful forecasting experience of the LDI model, the LDI price elasticity of −0.4 to −0.5 has to be taken seriously. However, as will be discussed below, most of the parameters in the LDI model are underidentified and have been estimated on the basis of simulation and expert judgment. Consequently, my own view at this point is that one can reasonably question whether the price elasticity in the LDI model may be too small (in absolute value).

As noted earlier, five toll demand studies have been selected for detailed review, namely, Deschamps, Larsen and McCleary, the LDI model of AT&T Long Lines, Pousette, and Waverman. The LDI model will be discussed first. Development of the LDI model was begun in April 1966, and an operational version was available by midsummer 1967.[43] The model is a computerized mathematical model that simulates the generation of domestic message toll service (MTS) messages, conversation minutes, and revenues. The market considered by the model is interstate MTS messages within the forty-eight contiguous states.

The LDI model simulates one month's (October) MTS message volume in great detail. The primary objective in developing the model was to provide management with a "flexible, consistent, accurate, and improvable system" for estimating the impact of a wide range of possible MTS rate schedules on the domestic interstate MTS market. Secondary objectives of the model included evaluating the impact of nonrate factors on MTS volumes and on the composition of volumes by time-of-day, type of traffic, and

---

[42] The price elasticity estimated by Wert is also fairly small. However, the dependent variable in Wert's analysis is the day time proportion of toll volume, and the price elasticity estimated is with respect to day/night price differentials.

[43] The full name of the model is the Long Lines' Long Distance Interstate MTS Demand Model. The development of the LDI model was carried on under the direction of Robert Auray, with major contributions from Kendall Murphy and Howard Whitmore. The discussion that follows is based upon the description of the LDI model presented in Betteridge (1973).

so on. However, the secondary objectives were given weight in developing the model only to the extent that they were consistent with the primary objective. The disaggregation in the model is as follows:

1. Three customer classes (business, residential, and coin);
2. Fifteen lengths-of-haul (starting with one to eight miles, then nine to thirteen, fourteen to eighteen, and so on, up to 2,301 to 3,000 miles;
3. Two types of call (station-to-station and person-to-person);
4. Three types of day (weekdays, Saturday, and Sunday);
5. Twenty-five times of day (twenty-three one-hour periods and two half-hour periods[44]).

The model is thus seen to include 6,750 separate sectors (or cells).

Each of the 6,750 cells is postulated to have a "message generator" equation of the form

$$\hat{M}_i = a_0 [X_{1i}^{a_1} X_{2i}^{a_2} X_{3i}^{a_3} X_{4i} a_6^{X_{6i}} a_8^{X_{8i}} X_{7i}] + X_{3i}^{a_3} \Sigma X_{60ij}. \tag{35}$$

In this equation,

> . . . predicted messages in a sector of the market for a given October (M_i) are initially defined as being equal to a multiplicative function of several variables. In the right-hand side of the equation, each of the X's represents the value of a variable for a certain October (i). The appropriate weighting for that variable in that sector of the market is indicated by the associated coefficient (a). In the equation, X_1 is the military variable, X_2 is the economic variable, X_3 is the price (stimulation) variable, X_4 is a composite of the competitive effects plus shocks variable, X_6 and X_8 are representations of time, and X_7 is the equivalent business days (or trading days) variable. The symbol a_0 represents messages in that sector of the market during a base period (October 1960). Observe, therefore, that the model defines estimated messages in period "i" in terms of *changes from the 1960 situation*, due to the fact that time, rates, economic conditions and other enumerated factors may be different than the values prevailing in 1960. All variables in October 1960 are defined such that in October 1960 their value is 1.000; that is, each variable is measued in ratio or index form, with October 1960 equal to 1.000. [Betteridge (1973), pp. 18–20, italics in original]

---

[44] The divided hour is 4:00–5 A.M.

The second term on the right-hand side in (36), $X_{3i}^{a_3}X_{60ij}$, represents the cross-elasticity, or the volume of messages "shifted or reclassified" from one cell to another when there is a change in price.[45] The summation in this term is over 150 separate cells, and reflects the fact that the model allows for shifting or reclassification from potentially 150 other sectors of the market into the given sector (literally only 149, excluding the given sector). The 150 sectors include the three groupings by type of day, two groupings by type of call, and twenty-five groupings by time of day. The possibility of shifting between mileage bands or class of customer is excluded.

The variable X_{60ij} thus represents the number of messages shifted from the jth cell to the ith cell as a result of a change in price in the ith or j^{th} cell. The sum ΣX_{60ij} then measures the total number of messages shifted into the ith cell from the 149 other cells. This total is then adjusted for price effects (i.e., multiplied by $X_{3i}^{a_{3i}}$) just as if the shifted messages had been originally included in the sector. The model thus assumes that when messages are shifted on balance from one sector to another, the customers shifting will generate additional calls (if the new cell has experienced a rate reduction) in the same proportion as those generated by customers already in that cell.

The value of x_{60ij} is determined according to the following "shift and reclassification" models:

$$X_{60ij} = [a_{0j}X_1^{a_{ij}}X_2^{a_{2j}}X_{3j}^{a_{3j}}X_4 a_{6j}^{X_6} a_{8j}^{X_8} X_7][X_{5ji}^{a_{5ji}} - 1]. \qquad (36)$$

The point of departure in the model is the number of messages that initially exist in cell *j*. This number is given by the first expression in brackets, and is simply the first term in the "message generator" [i.e., equation (35)] for cell *j*. The second expression in brackets measures the proportion of cell *j*'s messages that are shifted to cell *i*. The variable X_{5ji} in this expression represents the ratio of the rate index (1960 = 100) in cell *j* to the rate index (1960 = 100) in cell *i*. If the rate indexes in the two cells are the same, X_{5ji} is equal to one, and there will be no shift of messages.

---

[45] As used by the LDI authors, "shift" means the rate-responsive transfer of calls from one hour of the day to another or from one day of the week to another, while "reclassification" denotes the transfer of calls from one type of traffic to another type (person-to-person to station-to-station, for example).

However, if the indexes differ, messages will shift between the two cells, with the proportion being shifted depending on the value of the cross-elasticity, a_{5ji}.

The cross-price elasticity between cells j and i is a composite quantity that is assumed to depend, among other things, on the distance in time between cells j and i and the "relative desirability" of calling at the time in cell i as oposed to the time in cell j. Specifically, it is assumed that

$$a_{5ji} = k_1 \frac{D_{ji}}{(1 + W_{ji})k_2}, \tag{37}$$

where

k_1 = propensity to shift *anywhere* from cell j
W_{ji} = distance in time between cells j and i
k_2 = scaling factor for $(1 + W_{ji})$
D_{ji} = relative desirability of calling at the time of cell i versus the time of cell j.

Before describing the procedure by which the parameters in equations (35) to (37) were adduced, several variables in equation (35) remain to be discussed. Economic activity (X_2) is measured either by real disposable personal income (DPI) from the National Income Accounts (used for residence and coin markets) or by a weighted average of DPI and the Index of Industrial Activity constructed by AT&T. Each of these variables is measured as a ratio to its underlying trend. The price variable (X_3) is measured as a weighted average toll price for messages in a given market sector, defined as the charge to the customer plus applicable federal excise tax. X_3 for period i is then expressed as the toll price for i divided by the toll price for October 1960.[46] Substitute telecommunications service (X_4) is represented by a composite variable that reflects the impact of WATS and private line offerings and nonrecurring "shocks" on business MTS volume.

The variables X_6 and X_8 represent the "basic growth" in a sector. Specifically, the coefficient a_6 represents the growth rate of the given sector of the market prior to 1960, and a_8 represents the

---

[46] X_3 is not deflated by the CPI. The effects of general inflation are reflected in a trend term.

growth rate of that sector after 1960.[47] The military variable, X_1, is included in order to avoid confusing the impact on message volumes of the Korean war manpower buildup with the impact of the concurrent surge in economic activity. Specifically,

> . . . when a substantial buildup in the armed forces occurs, it creates a new demand for telephone communication which did not previously exist. Large numbers of men are moved to locations remote from their former residences, leaving behind a number of relatives and friends with whom communication is desirable. [Betteridge (1973, p. 12)]

Estimates of the number of messages for a sector from equation (35) are converted to revenue by multiplying by the average price of a call in the sector. However, since most rate structures involve a fixed charge for a specified initial period of time plus an additional charge for each subsequent period of the call, the average price of a call clearly depends on the distribution of calls with respect to duration. In general, these distributions vary over time, and this variation has to be taken into account. Specifically, it is assumed that the distribution of calls by length of conversation is lognormal. The parameters of the distribution—namely, the mean and standard deviation—were then estimated as coefficients in a least-squares regression equation using data for October. Equations were estimated for each sector for each of the Octobers for which data were available. From these, it was possible to estimate how the length-of-conversation distributions vary over time.[48]

The length-of-conversation distributions are used to calculate for each sector, the weighted average number of overtime periods for the "typical" call in the sector. This number is multiplied by the appropriate rate per overtime period, and to this is added the initial fixed charge. The resulting average price of a message in a sector is then multiplied by the number of messages for the sector estimated from equation (35) to yield an estimate of the revenue that is generated in the sector.

---

[47] This "split trend" approach is included in order to reflect a noticeable change in growth rates, particularly between station and person traffic, which occurred around 1960 in most sectors of the market. [See Betteridge (1973, pp. 20-21).]

[48] The results indicated that the mean lengths of calls vary with time, but that standard deviations do not. Trend rates of change in sector mean lengths of conversation were calculated from the estimated October means, and these were used in extrapolating the sector distributions forward or backward in time, taking October 1965 as the base.

Because of the large number of cells, values for the parameters in equations (35) to (37) had to be obtained in stages. Initial estimates of the parameters in equation (35) were derived by conventional multiple regression methods using aggregate time-series data for Octobers over the period 1948–1965. The coefficients obtained from these regressions were then used as benchmarks in a series of simulations of the disaggregated data, from which estimates of "shift and reclassification" effects were obtained. In this stage, time-series data which had been aggregated in the first stage were once again disaggregated into their original components (by type of call and rate period). The disaggregated model was then "calibrated" in a simulation framework. The coefficients estimated in the first stage served as points of departure in this calibration. Deviations of the actual number of messages from their simulated values were taken as estimates of shift and reclassification. From these estimates of shift and reclassification for 1963 and 1965, initial estimates of the parameters in equations (36) and (29) were derived.[49] Repeated simulation, with the residuals from a just completed simulation being used to modify the parameters used in the next simulation, eventually yielded the "final" model.

The parameters estimates obtained by this procedure can be briefly summarized as follows:[50]

1. Price elasticity coefficients [a_3 in equation (35)] range from about -0.10 to about -0.70. Small negative values are typical for short-haul, low-rate traffic, with progressively larger negative values being found for longer lengths of haul. Another generally consistent pattern is that the price elasticities for daytime traffic are noticeably smaller than for evening traffic.
2. The coefficients for economic activity [a_2 in equation (35)] vary from close to zero to as high as 2.5. Low values are associated with short-haul traffic, and higher values are generally associated with longer haul traffic.

---

[49] "This was done initially at high levels of aggregation (e.g., daytime traffic versus nighttime traffic; or station-to-station traffic versus person-to-person traffic) to obtain estimates of total shift/reclassification. Then the data were further disaggregated to allocate and anlyze the distribution of the shift by successively smaller and smaller time periods. Using the resulting detailed distribution or allocation of shifted and reclassified messages based on 1963 and 1965 experience, repeated simulations were made using the shift/reclassification model (see Charts 5 and 6) to develop coefficients for this model which approximated the actual distributions of shift and reclassification" [Betteridge (1973, pp. 41–42)].

[50] See Betteridge (1973, pp. 42–45).

3. A wide range of results were found for the "pure" growth rates [a_6 and a_8 in equation (35)]. With an average growth of about 10 percent per year for the period over which the model was calibrated, individual sector growth rates vary from -5 percent per year to over 30 percent per year. Again, there is a noticeable tendency for the coefficients to increase with length of haul. Growth rates of business and residence sectors are roughly equal, but both are consistently higher, other things being equal, than those for coin traffic.
4. The coefficient a_5 in the shift and reclassification model [equation (36)] is determined by the coefficients in equation (37). The "propensity to shift" coefficient [k_1 in equation (37)] varies between -0.002 and -0.65. *Ceteris paribus*, business k_1 coefficients tend to be smaller (in absolute value) than corresponding resident k_1's, and the latter tend to be lower than those for coin customers. Moreover, the absolute value of k_1 tends to increase with length of haul—that is, the longer the haul (and the larger the cost of the call), the greater is the propensity to shift the call. There does not appear to be any difference between the propensity to shift (i.e., to change the time of calling but not the class of call) and the propensity to reclassify from person-to-person to station-to-station.
5. The final coefficient determining the value of a_5 is k_2, the scaling factor for "waiting time" in equation (37). The values for k_2 range from 0.6 to 1.2. In general, the value of k_2 decreases as length of haul increases, and, other things being constant, are larger for business than for either residence or coin.[51]

As mentioned earlier, the LDI model is the most detailed model of toll demand in existence, and, according to its proprietors at Long Lines, it has fulfilled its mission well. After becoming operational, the model quickly came to be relied on by Long Lines and AT&T management as a tool of internal analysis and planning. Also, the model has been used in rate filings with the FCC, as well as in a number of intrastate toll filings before state regulatory commissions. The simulation framework of the LDI model allows it in certain circumstances to be easily modified without throwing the model out of balance, and this has been done occasionally in order to take into account breaks in underlying structure or to

---

[51] The LDI model authors refer to k_2 as the "propensity not to wait," since the higher its value, the lower the value of a_5. Thus, *ceteris paribus*, the larger the propensity not to wait, the lower the propensity to shift or reclassify.

adjust parameters in sectors in which the model is not tracking well.[52]

However, the strengths of the LDI model—namely, immense detail and a simulation framework that allows parameters to be modified without difficulty—are also its weaknesses. The detail available in the LDI model provides temptation for the model to be employed in circumstances for which it was not designed. Its use in intrastate toll filings are cases in point. The parameters of the LDI model were calibrated to interstate toll data for the continental United States. The LDI model is therefore a model of interstate toll demand for the entire continental United States, rather than being a model of intrastate toll demand for a specific state. And, while it is unlikely (at least in my opinion) that, *ceteris paribus*, intrastate and interstate toll demands are markedly different, toll demand per se may very well differ between states or regions of the country.

If this is the case, then the structure reflected in the LDI model represents, in effect, a mean for the forty-eight contiguous states as a whole, but, in general, will not mirror the toll demand behavior for any particular state.[53] That this is a real possibility renders the LDI model vulnerable in intrastate toll findings. The LDI model is also vulnerable in regulatory proceedings because of its simulation framework. Since the parameters in the model are calibrated by trial and error, rather than being estimated by econometric techniques, the model is open to charges that it is ad hoc and unscientific.[54] However, I want to emphasize that these remarks are not directed at the integrity of the model. Given the disaggregation that was desired, there was no way to obtain values of the parameters other than the technique that was used, since the data available could not support estimation by econometric methods.[55] Still, science and the realities of the marketplace require that the LDI model eventually be replaced by a model whose parameter values are arrived at by more objective procedures. Work is currently underway to do so.[56]

---

[52] See Betteridge (1973, pp. 47-49).

[53] That toll demand is honogeneous across states or regions is a hypothesis that needs testing.

[54] Unscientific in the sense that the parameter values finally arrived at are probably not reproducible, even by the same investigators.

[55] Calibration of the LDI model illustrates in an instructive way how expert judgment and limited data can join in estimating underidentified parameters.

[56] A description of this work currently under way is described briefly in Chapter 6.

Next on the list of models to discuss is Deschamps (1974). Deschamps's focus is on the demand for toll calls in Belgium during the period 1961–1969. The data employed consist of observations on toll traffic within and between districts for forty districts for one day for each of nine years. Deschamps's data set thus contains 14,400 observations in all. Deschamps estimated his models in the variance-components framework pioneered by Balestra and Nerlove (1966). The dependent variable in the analysis is the number of calls between district i and district j at time t. The logarithm of this quantity is then related to[57]

D_k, k=1, . . ., 10: a dummy taking the value of 1 if the distance in kms between i and j lines in the interval $(10(k-1), 20k)$;[58]

N_{it}, N_{jt}: the logarithm of the number of subscribers in district i or j at time t;

R_{it}: the logarithm of the income per capita in district i at time t;

P_{ij}: an index of sociological proximity, defined as the sum of the number of commuters from i to j and from j to i, and dividing by the aggregated population of districts i and j;[59]

L_{ij}: a dummy variable taking the value 1 if districts i and j speak the same language, 0 otherwise;

C_{ij}: a dummy variable taking the value 1 if district j contains the capital of the province to which zone i belongs in major part, 0 otherwise;

t: a time trend;

T_k, k=1, . . ., 9: a dummy variable taking the value 1 in sample year $60 + k$, 0 otherwise;

Q_{ij}: a dummy variable taking the value 1 if $i = j$, 0 otherwise;

$\overline{Q}_{ij}$: a dummy variable taking the value 1 if $i \neq j$, 0 otherwise.

The price variables considered by Deschamps were defined as follows:[60]

P^1_{ijt}: lump price per call, where applicable, 0 otherwise;

---

[57] See Deschamps (1974, pp. 12–13).

[58] Distance is calculated as the distance between the towns (or cities) giving names to the districts.

[59] This matrix was computed from data in the 1961 Census.

[60] See Deschamps (1974, p. 13).

P^2_{ijt}: the price of an additional minute beyond the first;
P^3_{ijt}: the minimum price per call, where applicable, 0 elsewhere;
P^4_{ijt}: $\max(0, P^3_{ijt} - P^2_{ijt})$;
P^5_{ijt}: the logarithm of a price index obtained by taking a weighted mean of the consumer costs of calls falling in different length classes (with weights independent of i, j, t);
P^6_{ijt}: the price of the first minute of call (all i, j, t).

In selecting these definitions of price, Deschamps is clearly cognizant of the need to allow explicitly for the fact that a telephone call is charged for on a two- or multi-part tariff.[61] The hypotheses considered by Deschamps are as follows:

1. One may assume that (a) that length of a call is somewhat beyond the caller's control once the connection is established, and that (b) the marginal utility of the next minute is not known with certainty by the caller. If this is so, it would seem appropriate to assume that the caller decides, before lifting the receiver, whether he will make the call or not; this on the basis of the mathematical expectation of the call's length, from which an expected cost can be derived. Quantities demanded here are numbers of calls. This leads to the inclusion of a *price index* (weighted mean) among the regressors.
2. One may instead conduct a traditional analysis in terms of the marginal utility of an additional minute and its marginal cost. This leads to a specification where the *price of an additional minute* appears as independent variable, along with the eventual *fixed part of the tariff.* The quality of service could also be introduced as explanatory variable, since an increase in the average number of attempts has formally the same effect as an increase in the connection charge. [Deschamps (1974, pp. 2–3), italics added.]

Deschamps examined eight different specifications of his model. For each specification, the 14,400 observations in the data set were "stacked," and then estimated by the variance-components technique of Balestra and Nerlove. The eight specifications considered by Deschamps involved alternative representations of toll price, distance, and time. The distance dummy variables, D_k, were included

[61] Cf. Chapter 2, Section 5.

in the model directly and also interacted with the number of subscribers in the district called. Time was represented by the linear trend, t, and the dummy variables, T_k. The best results are obtained with distance interacted with the number of subscribers in the called district and the time dummy variables.

Deschamps tested five different price formulations, namely:

(I) P^1, P^2, and P^3

(II) P^1, P^2, and P^4

(III) P^2 and P^6

(IV) $Q_{ij}P^6$, $\overline{Q}_{ij}P^6$, and P^2

(V) P^5.

Formulations I–III reflect the multi-part tariff pricing of a telephone call; of these, specification II is most in keeping with the consideration expressed in Deschamps' second hypothesis (and with my own discussion in Section 5 of Chapter 2). Use of the price index in specification IV, in contrast, reflects the view expressed in Deschamps' first hypothesis. Finally, specification V is intended as a limited test of the hypothesis that the price elasticity of demand varies with distance.

The results obtained by Deschamps can be briefly summarized as follows:[62]

1. Specifications I and II are practically equivalent. Both yield better results, although not drastically so, than specification III. "The price the consumer has to pay for the first minute indeed appears to be the major and most reliable determinant of the number of calls. This can be explained by the high percentage of short calls, a phenomenon apparent even when the price of an additional minute is zero."
2. The sign for the coefficient of $Q_{ij}P^6$ in specification IV is positive. Deschamps notes that this variable relates to intradistrict calls, varies almost exclusively with duration, and therefore probably does not display sufficient independent variation to identify its coefficient.[63]

---

[62] See Deschamps (1974, pp. 15-16).

[63] It is not clear why Deschamps did not test some models in which price was interacted directly with distance (unless, of course, distance and price are completely collinear).

3. Specification V gives the best results of all.[64] However, as Deschamps mentions, this may be due, at least in part, to the fact that p^5 is included as a logarithm.

Deschamps' results for specifications II and IV are presented in Table 3-6. In both of these equations, distance is represented by the distance dummy variable directly, rather than interacted with the number of subscribers in the district called. Also, time is represented by the linear time trend, t, instead of by the time dummy variables, T_k. Finally, it will be noted that income is absent altogether. In the models in which income is included, the sign is always negative

Table 3-6. Two Equations from Deschamps's Study of Demand for Toll Calls in Belgium[a]

Independent Variable	*Specification II*		*Specification IV*	
	Coefficient	*t-Ratio*	*Coefficient*	*t-Ratio*
constant	−3.62	−20.49	−3.51	−20.95
D_2	−0.98	−14.54	−0.92	−15.01
D_3	−1.64	−23.03	−1.52	−23.34
D_4	−1.97	−27.57	−1.84	−28.06
D_5	−2.18	−30.43	−2.05	−31.15
D_6	−2.22	−30.82	−2.10	−31.58
D_7	−2.34	−31.99	−2.21	−32.93
D_8	−2.38	−32.19	−2.25	−33.19
D_9	−2.40	−31.62	−2.27	−32.69
D_{10}	−2.42	−32.16	−2.29	−33.25
N_{it}	0.75	19.29	0.75	20.36
N_{jt}	0.99	75.18	0.99	83.49
t	−0.0073	−4.56	−0.097	−8.81
L_{ij}	0.38	27.81	0.38	30.64
C_{ij}	0.047	1.16	0.41	1.13
P_{ij}	1.58	6.67	1.47	6.89
p^1	−0.0031	−3.10	—	—
p^2	−0.0047	−1.68	—	—
p^4	−0.0053	−1.83	—	—
p^5	—	—	−0.24	−9.05
R^2	0.643		0.680	

[a]The dependent variable in both equations is the logarithm of the number of calls from district i to district j. The independent variables are as defined in the text.

---

[64] The price elasticity from this equation is the one tabulated in Table 3-5.

and statistically insignificant. Hence, income was excluded from these two specifications.

The most important predictors, it will be noted, are the dummy variables for distance, the number of subscribers, and whether or not the districts in question have a common language. The "sociological proximity" index (P_{ij}) is also highly significant. This variable and the dummy variable for common language (L_{ij}) are clearly very closely associated with the "community of interest" of the callers. Note, also, that the elasticity with respect to the number of subscribers in the calling district is smaller than the elasticity of the number of subscribers in the district called (0.75 versus 0.99). Deschamps mentions that this may reflect the presence of coin telephones, since calls can be made at coin stations, but, in general, are not received at them.

Although the price variables in specification II are not as strong statistically as the distance, system size, or community-of-interest variables, signs are right, and the *t*-ratio for P^1, which represents the fixed charge for the first minute is greater than 3 (in absolute value). In specification IV, in which toll price is represented by a price index, price is highly significant. But Deschamps notes that it may be that the superior performance of P^5 vis-à-vis P^1, P^2, and P^4 reflects nothing more than the fact that P^5 is included as a logarithm. Clearly, it would have been of interest to have estimated specification II with P^1, P^2, and P^4 in appropriate logarithmic form.

We now turn to the study of Larsen and McCleary (1970), which focuses on point-to-point demand for interstate toll calls. Larsen and McCleary would appear to be the first to model interstate toll demand on a point-to-point basis. Their study predates the study of Deschamps, which we just discussed and which also focuses on point-to-point traffic, by several years, and, in all likelihood, was unknown to Deschamps.[65]

Larsen and McCleary are concerned with explaining the volume of toll traffic flowing between individual pairs of states in the continental United States. Their procedure is to relate the volume of toll calls between state i and state j to income, price, and variables representing the "community of interest" between the states. A number of variables were considered as measures of community of interest, but the quantity receiving the most attention is the volume of mail (first-class and air mail) from state i to state j.

---

[65] Discussions with people at Bell Canada reveal that point-to-point toll demand modeling has also been undertaken at Bell Canada.

The model postulated by Larsen and McCleary has the form

$$1nV_{ij} = \mu + \alpha 1nM_{ij} + \beta 1nP_{ij} + \gamma 1nY_i + u_{ij}, \tag{38}$$

where

V_{ij} = the number of toll calls from state i to state j
M_{ij} = mail volume (first-class and air) from state i to state j
P_{ij} = the price of a call from state i to state j
Y_i = income per capita in state i
u_{ij} = random error term.

The data used in estimation consists of observations on directional traffic for all distinct state pairs for the forty-eight contiguous states and the District of Columbia. The model is estimated for business and residential traffic separately, as well as for total traffic consisting of business plus residence plus coin. Data were available for two four-month periods, October 1966 through January 1967 and March through June 1968.[66] Two definitions of price are considered, the average *initial* charge for a call between state i and state j and the average *total* charge for a call between the two states.[67]

The major findings of Larsen and McCleary are as follows:

1. The equations for business and residence customers are very similiar, although the fit for residence is slightly inferior to that for business.
2. Mail volume is the most significant predictor. For all practical purposes, toll volume and mail volume are proportional.
3. The coefficient of income is negative and significantly different from zero. The authors are unable to offer any explanation for this apparent anomaly.
4. Price is also highly significant statistically. Average total charge yields smaller estimated elasticities and somewhat lower R^2's than average initial price. With average initial price, the estimated price elasticity is in the neighborhood of -1.
5. There appears to be some inhomogeneity between bordering state pairs and nonbordering state pairs. The model appears

[66] Estimates of V_{ij} for these two periods were obtained from the Message Analysis Sampling Plan (MASP).

[67] See Larsen and McCleary (1970, pp. 10–12) for definitions of these variables.

to be most applicable to bordering state pairs. Washington, D.C., because of the large number of government calls, is a definite outlier.

The following equation for total toll traffic using October 1966 data on all state pairs and average initial price typifies the results obtained by Larsen and McCleary:

$$\underset{}{ln V_{ij}} = \underset{(95.71)}{2.01} + \underset{(127.5)}{1.02\, lnM_{ij}} - \underset{(-25.95)}{1.09\, lnP_{ij}} \qquad R^2 = 0.912. \tag{39}$$

Although it is not referred to as such, it is clear that the model used by Larsen and McCleary can be interpreted as a gravity model. "Mass" is measured by the volume of mail and income (although income has the wrong sign) and "distance" by price. Noteworthy, too, is the emphasis that is given to the community of interest of the calling parties. Larsen and McCleary measure community of interest by mail volume, and, given its importance as a predictor [cf. the *t*-ratio of *lnM* in equation (39)], the authors must be applauded for their choice.

All in all, the Larsen-McCleary study must be considered a good first effort at modeling interstate toll demand on a point-to-point basis, and it is unfortunate that it has not received a wider readership. Nevertheless, the study is not without problems. The perverse importance of income clearly requires explanation, as does also (at least in my opinion) the weak showing of the number of telephones.[68] Preferably, too, toll price should be represented by the average charge for overtime periods, as well as by the average initial charge. Finally, in using state pairs as the unit of observation, the Larsen-McCleary analysis is at a very high level of aggregation. Among other things, the notion of community of interest is clearly much more ambiguous for a state than for a city or town.[69]

---

[68] The culprit in both cases may be mail volume in that mail volume may be a good proxy for income and system size, as well as being an excellent measure of community of interest. Larsen and McCleary do not include the correlation matrix of the variables, so it is not possible to deduce the extent of multicollinearity among the predictors.

[69] Paul Brandon of Bell Laboratories has pointed out that there will be a bias in the estimate of the price elasticity in the Larsen-McCleary study if there is a residual effect of distance that is not reflected in the mail volume. Both the probability of sending a letter and the probability of making a call will decrease with distance. However, if these probabilities do not decrease at the same rate, there will be a residual distance effect that, since the price of a call increases monotonically with distance, will be reflected in the estimate of the price elasticity.

The final study to be discussed in this section is Waverman (1974). Waverman estimates equations for toll (or trunk) demand in Canada, Great Britain, and Sweden, using dynamic models of the Houthakker-Taylor variety. Time-series data are utilized in all cases. For Canada, equations are estimated for both business and residence demand using data for Bell Canada. The dependent variable in each case is deflated toll revenue. For Great Britain, equations are estimated for total inland trunk demand for deflated toll revenue and the number of calls as dependent variables. The dependent variable in the equations for Sweden is the number of trunk "pulses," and alternatively the number of trunk calls, for business and residential customers combined. For Great Britain, Sweden, and residential demand for Bell Canada, equations are estimated both on a per-capita and per-telephone basis. For business demand for Bell Canada, equations are estimated for deflated toll revenue and deflated toll revenue per telephone.

The following equations illustrate Waverman's results:[70]

CANADA

Residence

$$\underset{}{ln(R/T)_t} = \underset{(-8.20)}{-4.65} + \underset{(4.35)}{1.03}\, lnX_t - \underset{(-8.01)}{1.16}\, lnP_t \qquad (40)$$

Business

$$ln(R/T)_t = \underset{(-3.21)}{-2.24} + \underset{(2.83)}{0.11}\; ln(R/T)_{t-1} + \underset{(3.67)}{0.45}\, lnX_t - \underset{(-3.53)}{1.20}\, lnP_t \qquad (41)$$

R = real toll revenue
T = number of telephones
X = real personal disposable income
P = average toll price/implicit deflator for GNP.

GREAT BRITAIN

$$(R/T)_t = \underset{(1.05)}{12.82} + \underset{(3.45)}{0.51}(R/T)_{t-1} + \underset{(2.06)}{0.0017}(X_t + X_{t-1})$$

$$- \underset{(-2.82)}{0.041}(P_t + P_{t-1}) - \underset{(-1.62)}{0.051}(d_t + d_{t-1}) \quad \overline{R}^2 = 0.986 \qquad (42)$$

---

[70] Waverman (1974, pp. 33, 36, 13, 43). The elasticities from these equations are tabulated in Table 5. Waverman does not tabulate R^2's for his equations for Bell Canada.

$$(C/T)_t = \underset{(1.84)}{83.50} + \underset{(2.58)}{0.56}(C/T)_{t-1} + \underset{(2.64)}{0.0064}(X_t + X_{t-1})$$

$$- \underset{(-3.59)}{0.15}(P_t + P_{t-1}) - \underset{(-2.34)}{0.30}(d_t + d_{t-1}) \qquad \overline{R}^2 = 0.997 \tag{43}$$

R = real revenue from inland truck calls
C = number of inland trunk calls
T = number of telephones
X = real personal consumption expenditures
P = average price for trunk call/implicit deflator for personal consumption expenditures
d = percentage of trunk calls dialed by operator.

SWEDEN

$$lnPU_t = \underset{(1.10)}{-3.86} + \underset{(4.91)}{0.53}\, lnPU_{t-1} + \underset{(0.52)}{0.23}\, lnX_t - \underset{(-4.10)}{0.51}\, lnP_t + \underset{(2.87)}{0.89}\, lnT_t$$

$$\overline{R}^2 = 0.998 \tag{44}$$

$$lnC_t = \underset{(-1.20)}{-3.56} + \underset{(3.80)}{0.51}\, lnC_{t-1} + \underset{(0.45)}{0.19}\, lnX_t - \underset{(-2.43)}{0.29}\, lnP_t^* + \underset{(3.63)}{1.20}\, lnT_t$$

$$\overline{R}^2 = 0.998 \tag{45}$$

PU = number of trunk "pulses"
C = number of trunk calls
X = real Gross Demestic Product per household
P = marginal pulse rate/consumer price index
P^* = average price of a trunk call/consumer price index.

Waverman's results for Sweden are especially interesting, for the elasticity with respect to the number of telephones is less than 1 in equation (44), but greater than 1 in equation (45). Since the dependent variable in equation (44) is the number of pulses (which measures holding time), while the dependent variable in (45) is the number of calls, and since a coefficient greater than 1 on the number of telephones can be interpreted as representing a positive externality associated with system size, this result is consistent with the view that the number of calls responds to this externality, but that duration of calls does not. Interesting, also, is the fact that the price elasticity in the equation for pulses is larger than in the equation for calls. What this states is that duration is more responsive to price than is the number of calls.

This last result is confirmed in Waverman's equations for Great Britain. In equation (39), the dependent variable is average real trunk-call revenue per telephone, while in equation (40), the dependent variable is the average number of trunk calls per telephone. Thus the average duration of a call is reflected in equation (42), but not in equation (43). In Table 3-5, we see that the price elasticities are larger in the equation explaining real revenue per telephone than in the equation explaining the number of calls per telephone. Consequently, once again we can conclude that the duration of calls is more responsive to a price change than the number of calls. Moreover, the same conclusion, it should be noted, also holds for the income elasticities. In both Great Britain and Sweden, the duration of long-distance calls is more responsive to income changes than the number of calls.

Chapter 4

The Empirical Literature on Telephone Demand: Intrastate Toll, WATS and Private Line, Coin, Etc.

TOLL DEMAND

Intrastate Toll Demand

Studies of intrastate message toll demand comprise a lion's share of the total literature on telephone demand. This attention reflects the increasingly common practice of utilizing an econometric model in estimating the impact on revenue of proposed changes in tariffs in rate filings before state regulatory commissions. The various state modeling efforts have, for the most part, employed the common framework of relating a physical measure of intrastate message toll activity to a measure of economic activity, the relative price of MTS calls, the size of the market, and "habit." In all cases, MTS activity is measured by the number of intrastate toll messages or by intrastate toll revenues deflated by an MTS price index. Economic activity is usually measured by real personal income or real disposable personal income, but other measures of economic activity are used as well. These include Gross State Product and bank debits. MTS price is usually measured by a Laspeyres price index that aggregates over mileage bands, time of day, and class of customer.[1] For two states, interstate rates are represented as well.

Market size is usually represented by the number of telephones less residence extensions or by the number of main stations, either explicitly as a predictor or implicitly as a deflator of toll activity.

[1] For more than a third of the states, a chain index is used instead of the Laspeyres index.

In the models for nine states, market size is measured (explicitly or implicitly) by population or the number of households. Finally, in all cases but one in which "habit" is represented, it is measured by the lagged value of the dependent variable.

As already noted, personal income (or disposable income) is expressed in real terms, usually on a per capita or per household basis. Where local deflators are not available, the national CPI is usually used as a surrogate. In nearly all cases, the CPI is used to deflate the toll price index, so that toll price is measured relative to the cost of other goods and services. Quarterly time-series data are used in estimation of the models, with one exception where monthly data are used.

Most of the intrastate toll models are variations of the following logarithmic Koyck distributed-lag model:

$$lnM_t = \alpha_0 + \alpha_1 \, lnM_{t-1} + \alpha_2 \, lnX_t + \alpha_3 \, lnP_t + \alpha_4 \, lnT_t + u_t, \tag{46}$$

where

M = number of intrastate toll messages in a state[2]
X = real personal income per capita in the state
P = toll price index for the state divided by the CPI
T = number of telephones in the state less the number of residence extensions
u = random error term.

Included among the variations of this model, one finds

1. Restricting α_4 to equal 1. In this case, the dependent variable is lnM/T.
2. Specifying the equation as linear rather than double logarithmic.
3. Restricting α_1 to equal 0, in which case lnM_{t-1} is omitted as a predictor.[3]

[2] Usually the number of toll messages is adjusted for conversion to extended area service (EAS) during the period of the sample. In some states, conversion to EAS has created major problems in defining a dependent variable that is consistent over time.

[3] The model may still be dynamic depending on the presence or absence of the number of telephones as a predictor. If the number of telephones is present, then the model is dynamic since the number of toll calls is *conditional* on the stock of telephones.

4. Use of a chain index for toll price rather than the Laspeyres index.
5. Including prices of substitute forms of communication as predictors.[4]
6. Finally, since quarterly (or monthly) data are used in estimation, some form of seasonal adjustment is required. Dummy variables are used in about half of the models, and the Shiskin (Census X-9 or X-11) method in the rest.

The price and income elasticities of demand from the state models are tabulated in Table 4-1.[5] As mentioned, the models are estimated from quarterly data except for one estimated from monthly data. Typically over ten years of data are used in estimation. Also, all of the models are multiplicative except for three which are linear. The lagged dependent variable is absent from four of the models.[6] However, of these, only one model is truly static. The other three are conditional on the stock of telephones, or include lagged price and income terms.

The estimates of the price elasticities in Table 4-1 vary from −0.03 to −0.44 in the short run and from −0.22 to −1.04 in the long run. The mean of the short-run estimates is −0.21, while the mean of the long-run estimates is −0.67. There is greater variation in the estimates of the income elasticities, but here, too, there is substantial agreement among the estimates. The estimates range from 0.05 to 1.08 in the short run and from 0.28 to 2.72 in the long run, with means of 0.39 and 1.33, respectively.[7]

Thus, the evidence on price and income elasticities from the intrastate toll demand models indicates

1. That intrastate toll demand is more responsive (measured in terms of elasticity) to changes in income than to changes in toll price;

[4] One model includes a price index for other telephone services; the first-class letter rate appears as a predictor in two others.

[5] I am grateful to the Demand Analysis Group, Tariffs and Costs Department, at AT&T for undertaking a detailed tabulation of the state models. Edward Dinan of New England Telephone worked on this study while a participant in the Internship Program sponsored by the Demand Analysis Group. The state models reviewed in this section were completed in 1976 through mid-1978, and represent, in many cases, the initial efforts of the Bell System operating telephone companies to build aggregate MTS demand models. Currently, there is considerable work underway updating and refining the existing models and developing new, more disaggregated models.

[6] The elasticities for these models appear in parentheses in Table 4-1 in order to emphasize that, since the models do not allow for habit formation, they will, in general, be intermediate between short-run and long-run elasticities.

[7] The estimates from the four models that do not allow for habit formation are excluded in calculating mean values of the price and income elasticities.

Table 4–1. Estimates of Price and Income Elasticities of Demand for Intrastate Toll Calls (Bell System)

State	Dependent Variable	Price Elasticity: Short-Run	Price Elasticity: Long-Run	Income Elasticity: Short-Run	Income Elasticity: Long-Run	Form of Model	Remarks
State A-1	M	(−0.16)		(0.92)		Linear	Total toll includes interstate; estimated by Durbin Two-Stage Method; chain toll price index
State A-2	M/MT	−0.15	−0.22	0.58	0.86	Log Koyck	CPI separate variable Chain toll price index
State A-3	M/MT	(−0.12)		(0.88)		Linear	Total toll includes interstate; estimated by Durbin Two-Stage Method; chain toll price index
State B-1	M/T	−0.32	−0.60	0.63	1.18	Log Koyck	Chain roll price index
State C-1	M	−0.07	−0.14	0.24	0.53	Log Koyck	Average number of telephones less residential extensions included as predictor
State D-1	PDR	−0.35	−0.45	0.31	0.40	Log Koyck	Chain toll price index time included as predictor
State E-1	M/MT	−0.03	−0.85	0.05	1.40	Log Fl-Adj	
State E-2	M/MT	−0.21	−0.73	0.19	0.65	Log Koyck	
State E-3	M/MT	−0.17	−1.04	0.09	0.55	Log Fl-Adj	
State E-4	M/T	−0.26	−1.04	0.14	0.56	Log Koyck	
State E-5	M/MT	−0.13	−0.81	0.21	1.28	Log Koyck	

State F-1	PDR/POP	−0.14	−0.62	0.21	0.91	Log Koyck	Main telephones per capita included as predictor
State G-1	PDR/POP	−0.16	−0.56	0.80	2.71	Log Koyck	Telephones less residential extensions per capita included as predictor
State H-1	PDR	−0.37	−0.50	0.63	0.88	Log Koyck	Time and price of other telephone services included as predictors; chain toll price index
State I-1	M/T	−0.44	−0.84	0.19	0.36	Log Koyck	Chain toll price index
State I-2	PDR/POP	−0.29	−0.64	0.24	0.54	Log Koyck	Chain toll price index; telephones less residential extensions per capita included as predictor
State I-3	M/T	−0.35	−0.96	0.31	0.81	Log Koyck	Chain toll price index
State I-4	M/T	(−0.59)		(0.52)		Double Log	Chain toll price index; estimated by Cochrane-Orcutt
State J-1	PDR/POP	−0.14	−0.23	0.24	0.58	Log Koyck	Telephones less residential extensions per capita included as predictor; also weather
State K-1	PDR/T	−0.21	−0.91	0.39	1.70	Log Koyck	
State L-1	M	−0.20	−0.39	0.45	0.85	Log Koyck	Households included as predictor
State L-2	M	−0.23	−0.43	0.59	1.12	Log Koyck	Households included as predictor
State M-1	PDR/POP	−0.12	−0.69	0.47	2.77	Log Koyck	First class postal rate included as predictor

Table 4–1 continued

State	*Dependent Variable*	*Price Elasticity*		*Income Elasticity*		*Form of Model*	*Remarks*
		Short-Run	*Long-Run*	*Short-Run*	*Long-Run*		
State M-2	PDR/POP	−0.17	−0.83	0.46	2.23	Log Koyck	First class postal rate included as predictor
State N-1	PDR/POP	−0.14	−0.82	0.33	2.01	Log Koyck	Change in insured unployment rate included as predictor
State N-2	PDR	−0.24	−0.86	0.60	2.10	Log Koyck	
State N-3	PDR/POP	−0.15	−0.79	0.44	2.35	Log Koyck	
State N-4	PDR/POP	−0.13	−0.91	0.36	2.52	Log Koyck	
State O-1	PDR/POP	−0.07	−0.84	0.23	2.65	Log Koyck	First order transformation; unemployment rate included as predictor
State R-1	PDR/POP	(−0.21)		(0.59)		Linear	Dependent variable first difference
State Q-1	PDR	−0.31	−0.37	1.08	1.32	Log Koyck	Dependent variable includes private line & WATS; second order GLS transformation; chain total toll price index

Symbols: M: messages; MT: main telephones; T: telephones less residential extensions; PDR: price-deflated revenues; POP: population; Fl-Adj: Houthakker-Taylor flow-adjustment model.

2. That intrastate toll demand is inelastic with respect to both price and income in the short run; and
3. That intrastate toll demand, broadly but not everywhere, remains inelastic with respect to price in the long run, but is elastic in the long run with respect to income.

It will be noted that toll price appears in all of the models in Table 4–1 and that income (or some other measure of economic activity) also appears in all of the models. Further, signs are correct in every case, and, in general, t-ratios are impressive (at least for time-series data). For the majority of models, the *t*-ratio for toll price is at least 3 (in absolute value). The smallest *t*-ratio for toll price is 0.99, and the next smallest is 1.99. For income, *t*-ratios are frequently 4 or more. The smallest *t*-ratio for income is 0.95, while the next smallest is 1.48.[8] Thus, it has to be concluded that both toll price and income are major determinants of intrastate toll traffic.

The number of telephones (main stations or total telephones less residence extensions) appears in eighteen of the state models, either as a deflator of the dependent variable (thirteen models) or as an independent variable (five models). In the generic model, the stock of telephones is included, as noted, as a measure of market size.[9] Naturally, we expect more intrastate toll calls to be made in a state with a larger number of telephones, and this is presumably what the term "market size" is meant to convey. In the thirteen equations in which the number of telephones appears only as a deflator of the dependent variable, the assumption, clearly, is that, *ceteris paribus*, toll activity increases at the same rate as the size of the system. However, in Chapter 2, we emphasized the externalities that are associated with the telephone. The access externality makes belonging to a larger system more valuable than belonging to a smaller system, and this should be reflected in more calls per telephone being made in a larger system than in a smaller system. It was suggested that the strength of the call externality could also be measured by system size. Thus, if access and call externalities are as just

---

[8] Interestingly, there are more cases of *t*-ratios less than 2 for income than for price (6 versus 2). The model for State E–1 has the smallest *t*-ratio for both price and income. These poor results almost certainly reflect data problems caused by extensive conversion to EAS in that state's largest metropolitan area in the 1960s and early 1970s. Because of this, it is difficult, if not impossible, to get a consistent time series for the number of toll calls.

[9] For States L–1 and L–2, market size is measured by the number of households.

described, the increase in toll activity will, *ceteris paribus*, be faster than the increase in system size.

However, the empirical evidence for these externalities is mild. Seven states have market-size variables included as predictors, and the coefficients of interest are tabulated in Table 4–2. The only states with the coefficient on the market-size variable greater than 1 are the two with market size measured by the number of households.

Some of the more individual features of the state toll demand models include

1. As already noted, the toll price index in two of the state models is a combined index that reflects interstate tariffs as well as intrastate. Implicit in these models is a "total bill" concept, which reasons that subscribers, especially residential subscribers, budget for interstate and intrastate toll calls jointly. This is an interesting idea, but it cannot be readily tested using aggregate data.[10]
2. In the models for nineteen states, the toll price index includes the Federal excise tax. While there does not appear to exist any difference in results that can be attributed to this adjustment, in principle, it is an appropriate adjustment to make. In five models, local sales taxes are included in addition to the federal excise tax.

Table 4–2. Coefficients of Market-Size Variable (Bell System Intrastate Toll Demand Models)

State	*Coefficient*	*t-Ratio*
Telephones as measure of market size		
State C-1[a]	0.93	3.90
State F-1[b]	0.46	2.08
State G-1[c]	0.53	2.96
State J-1[c]	0.97	3.03
State J-2[c]	0.92	2.67
Households as measure of market size		
State L-1	1.29	7.91
State L-2	1.39	7.59

[a]Total telephones less residence extensions.
[b]Main stations per capita.
[c]Total telephones less residence extensions per capita.

[10]For a discussion of the "total bill" concept applied to all telephone related expenditures, see Doherty and Oscar (1977). Alternatively, one can interpret the use of a combined index as a reflection of cusumers confusing intrastate and interstate tariffs.

3. Although personal income (or disposable personal income) is typically used as the measure of economic activity, there is some variation across the states. Personal income is augmented by changes in the rate of insured unemployment in the models for States N-1, N-2, N-3, N-4, and O-1. Gross State Product is used in place of personal income in the model for State A-1. Finally, bank debits is used in lieu of income in the model for State I-4.[11]
4. Three state models include price variables in addition to toll price and the CPI (or other general deflator). State H-1 includes a price index (deflated by the CPI) reflecting the costs of basic service, additional message units, and interstate toll. In this case, inclusion of the costs of these other services is derived from the "total bill" concept. The sign of the variable is negative, in line with the hypothesis advanced, and its *t*-ratio is −1.97. The additional price variable in models for States M-1 and M-2 is the first-class letter rate (also deflated by the CPI). Its sign is positive, as expected, but its *t*-ratio is only 1.76 and 1.42, respectively.
5. A final item of interest is the finding in the model for State J-1 that toll activity is sensitive to inclement weather.

Thus far, the focus in this section has been on models of individual states developed by Bell System operating companies. In all cases, the models refer to residence and business MTS messages combined and are estimated with time-series data. We now turn to two other analyses of intrastate toll, one the cross-sectional study of Feldman (1976) and the other a combined time-series/cross-sectional study conducted by Stuntebeck (1976).

For both business and residence subscribers, the Feldman study (which was also discussed in connection with the demand for access in Section 2 of Chapter 3) specifies models for the following categories of MTS demand:

Number of Messages
- Operator-Handled
- Person-to-Person
- DDD

[11] This substitution was prompted by the erratic behavior in the personal income series occasioned by sharp swings in agricultural prices. In this connection, it is interesting to note that measures of "permanent" income, constructed from moving averages of personal income, are used in the models for States B-1, D-1, I-3, and Q-1.

Number of Messages/Main Station
- Operator-Handled
- Person-to-Person
- DDD

Number of Minutes/Message
- Operator-Handled
- Person-to-Person
- DDD

In the equations with messages as the dependent variable, the independent variables are toll price, the price of alternatives, price of local service, population (Bell-served area), per-capita personal income, a proxy for the size of the exchange area,[12] and regional dummy variables.[13] In the equations for average length of conversation, the independent variables are toll price, average length-of-haul, per-capita personal income, and regional dummy variables. The equations, double-logarithmic in all cases, are estimated for business and residence subscribers separately, using cross-sectional data for the fourth quarter of 1973.

The price and income elasticities obtained in the eighteen equations listed above are tabulated in Table 4-3. In general, the price elasticities in the equations for messages are consistent with those obtained in the state studies, but this is not the case for the income elasticities. Income is absent from all but one of the residence message equations, and from four of the business message equations. In contrast, income appears in all of the equations for average length of conversation.

Feldman's results will be illustrated in detail with the equations for messages per main station and average length of conversation for DDD calls by residential customers. The equations are as follows (t-ratios are in parentheses):

$$\begin{aligned} ln(\widehat{M/T}) = \underset{(9.81)}{4.85} - \underset{(-4.31)}{0.58}\, lnP_{DDD} - \underset{(-1.02)}{0.18}\, ln\pi_L + 0.04\, ln(P_{OP}/P_{DDD}) \\ + 0.02\, ln(P_{PER}/P_{DDD}) + \underset{(5.94)}{0.13}\, lnPOP - \underset{(-10.94)}{0.62}\, lnEA \\ + \text{regional dummy variables} \qquad \bar{R}^2 = 0.88 \qquad (47) \end{aligned}$$

---

[12]This variable is intended to take into account variation in exchange areas across states. It is measured by the ratio of local calls to intrastate toll calls. Feldman also experimented with a "call-preference" proxy, defined as the ratio of intrastate toll calls to interstate calls, as a measure of community of interest. However, multicollinearity prevented the use of both variables as predictors at the same time.

[13]The average length-of-haul is an additional predictor in a few of the message equations.

Table 4–3. Estimates of Price and Income Elasticities for Intrastate Toll Demand (Feldman Study)

Dependent Variable	*Residence*		*Business*	
	Price Elasticity	*Income Elasticity*	*Price Elasticity*	*Income Elasticity*
Messages				
Station-to-station (operator-handled)	−0.31	—	−0.38	1.02
Person-to-person	−0.75	—	−1.31	—
DDD	−0.53	0.20	−0.49	0.74
Messages/Main Station				
Station-to-station (operator-handled)	−0.40	—	−0.46	—
Person-to-person	−0.70	—	−0.58	—
DDD	−0.64	—	−0.53	—
Minutes/Message				
Station-to-station (operator-handled)	−0.004	0.24	—	0.19
Person-to-person	−0.17	0.35	−0.15	0.06
DDD	−0.14	0.25	—	0.10

Source: Feldman (1976).

$$\ln \hat{ALOC} = \underset{(3.01)}{0.60} - \underset{(-2.45)}{0.14} \ln P_{OP} + \underset{(4.24)}{0.13} \ln ALOH + \underset{(4.37)}{0.25}(PI/POP)$$

$$+ \text{ regional dummy variables} \qquad \bar{R}^2 = 0.69, \qquad (48)$$

where

M = number of residence DDD intrastate MTS messages
T = number of residence main stations
π_L = local revenue per main station (residence)
POP = population
EA = local messages/intrastate MTS message (residence)
$ALOC$ = average length of conversation (residence DDD MTS messages)
P_{DDD} = average revenue per minute (residence DDD MTS messages)
P_{OP} = average revenue per minute (residence operator-handled messages)
P_{PER} = average revenue per minute (residence person-to-person messages)
$ALOH$ = average length of haul (residence DDD MTS messages)
PI = personal income.

In the equation for DDD messages per main station, it is interesting to note the importance of toll price (measured by average revenue per message),[14] population, and the exchange area proxy. In general, the relative order of magnitude of the estimated price elasticities in the residence messages and messages per main station in Table 4–3 are exactly what we should expect: the price elasticity for station-to-station calls is smaller than for DDD calls, which is smaller than for person-to-person calls. The negative coefficient for the exchange area proxy, with its large t-ratio, suggests an inverse relation between the sizes of local and "foreign" communities of interest.[15] In the equation for average length of conversation, we find, as expected, a positive (and highly significant) coefficient

---

[14] The estimates of 0.04 and 0.02 for the cross-price elasticities are taken from the equations for residence operator-handled and person-to-person messages, respectively. This is done in order to impose consistency in the cross-price elasticities across equations.

[15] However, one must be somewhat dubious of this result because of the fact that the sum of total MTS and WATS messages appears in the denominator of EA. This can set up a spurious negative relationship, and the substantial t-ratio for EA may be a reflection of this. Still, the idea that there is a trade-off between local and "foreign" communities of interest is worthy of further study.

for the average length-of-haul. Price (Now measured by average revenue per minute) continues to be important, although it is not as strong as in the equation for messages per main station. Note, however, that personal income, which is absent altogether from the message equation, is the strongest predictor in the duration equation.

The focus in Stuntebeck's study is on the relationship between day/night message distributions and day/night rate structures for residence weekday intrastate toll calls. A model is specified that relates the ratio of daytime to nighttime message volume in a mileage band to the ratio of the daytime price to the nighttime price,[16] personal income per capita, percentage of the population under eighteen years of age, percentage of the population over sixty-five,[17] and the average distance of a call in the mileage band. A double-logarithmic equation is estimated for three mileage bands, 25-50, 51-100, and 101-200 miles. State cross-sectional data for October-December 1968, March and April 1969, October-December 1970, March and April 1971, October 1972, and March 1973 are used in estimation.[18] A set of equations is estimated using the data for October-December 1970, and another set is estimated pooling the observations for three of the six years.[19]

The price and income elasticities estimated by Stuntebeck in the cross-section equations for October-December 1970 are tabulated in Table 4-4. Composite elasticities, obtained as weighted averages

Table 4-4. Estimates of Day/Night Price and Income Elasticities (Stuntebeck Study, Cross-Section Sample)

Mileage Band	*Price Elasticity*	*Income Elasticity*
25-50 miles	−0.11	0.56
51-100 miles	−0.27	0.58
101-200 miles	−0.59	0.18
Weighted mean	−0.21	0.52

Source: Stuntebeck (1976, p. 20).

[16] Daytime is counted as 8:00 A.M. through 4:59 P.M., and nighttime as 5:00 P.M. through 7:59 A.M.

[17] The age variables are included to reflect "daytime opportunity."

[18] The data for 1969-1971 are the Message Analysis Sampling Plan (MASP), while the data for 1972-1973 are from the Centralized Message Data System (CMDS).

[19] In the pooled equation, the population and distance variables are replaced with state dummy variables.

(using message volumes as weights) of the elasticities across separate mileage bands, are also presented. The price elasticities are seen to increase (ignoring sign) with length-of-haul; the income elasticities, in contrast, are seen to decrease. The weighted mean elasticities for price and income are −0.21 and 0.52, respectively.

Since the price income elasticities being estimated refer to the impact on relative day/night volumes of changes in day/night price differentials, these results clearly suggest that long-haul calls are shifted more than short-haul calls. Given that long-haul calls are also longer in duration than short-haul calls, this result is probably to be expected since the charge for a long-haul call increases more than in proportion to the increase in distance. However, an explanation for the sharp decrease in the income elasticity after one hundred miles is not readily apparent.[20]

The results from the pooled cross-section/time-series equations are presented in Table 4-5. Two equations are estimated for each mileage band, the observations from 1968, 1970, and 1972 being used in one equation, and 1969, 1971, and 1973 being used in the other. The years are split in order to account for seasonality. Once again, price elasticities are seen to increase with length-of-haul, while income elasticities decrease. In this case, though, the price elasticities are uniformly smaller, and income elasticities uniformly larger, than the estimated pure cross-section elasticities in Table 4-4.

A lot of material has been covered in this section, and it will be useful to summarize what I consider to be the major findings. Although there is considerable variation in the estimated price and

Table 4-5. Estimates of Day/Night Price and Income Elasticities (Stuntebeck Study, Pooled Sample)

Mileage Band	*Price Elasticity*	*Income Elasticity*
25-50 miles		
1968/70/72	−0.05	1.21
1969/71/73	−0.10	1.24
51-100 miles		
1968/70/72	−0.21	0.79
1969/71/73	−0.22	1.08
101-200 miles		
1968/70/72	−0.24	0.72
1969/71/73	−0.24	0.82
Weighted mean	−0.13	1.10

Source: Stuntebeck (1976, p. 32).

[20] Carl Pavarini has suggested that persons with high incomes may make proportionately more long-haul calls.

income elasticities from the state models in Table 4–1, there are also strong currents of uniformity. Intrastate toll demand is strongly influenced by the relative price of toll and income. The mean estimates of the long-run elasticities are −0.67 and 1.33 for price and income, respectively. The estimates of the price elasticities for interstate toll are, in general, smaller than the estimates, presented in Table 3–4 for interstate (i.e., long-haul) toll, but since the average length-of-haul is longer for interstate calls than for intrastate calls, this is to be expected. On the other hand, the intrastate toll elasticities are, again, in general larger than the elasticities that have been estimated for local use. This, too, is to be expected. Thus, the pattern that has emerged is: The elasticity for the service-connection charge is smaller than the elasticity for the monthly service charge, which is smaller than the elasticity for the intrastate toll price, which is smaller than the elasticity for the interstate toll price. And in Section 6 below, we shall see that the price elasticity for the interstate toll price is, in general, smaller than the elasticity for the price of international calls.[21]

WATS AND PRIVATE LINE

We now turn to the demand for WATS and private-line services. WATS, which can be viewed as an extension of flat-rate service to long-distance calling, is of two types: inward WATS and outward WATS. With inward WATS, the called party pays for the service, while with outward WATS, the caller pays. With either type, calls are routed through special designated WATS-servicing central offices. With private line, however, telephones are connected via dedicated direct circuits, rather than through exchanges. In light of the discussion in Chapter 2, it is clear that the demand for inward WATS will be closely associated with the call externality imposed on callees. Willingness to pay in this case is identified with the amount that a subscriber is willing to pay in order to receive calls. The demand for outward WATS, in contrast, is related to more traditional benefit/cost calculations. A cost-minimizing subscriber will choose WATS over MTS and private line only if the anticipated volume of calls at the WATS rates is cheaper than paying MTS and private-line rates.[22]

The literature on the demand for WATS and private line is small.

[21] The reference here is, of course, to the absolute values of the price elasticities.

I have found only three studies that focus explicitly on these services, namely

1. Davis et al. (1963), who estimate demand equations for WATS (inward and outward combined) and private line, using quarterly time-series interstate data for the United States.
2. Feldman (1976), who estimates equations for intrastate inward WATS and outward WATS separately, using cross-sectional state data for the United States.
3. Subissati (1973), who estimates equations for WATS (inward and outward combined) and business toll, using monthly time-series data for Bell Canada.

Of these three studies, the Feldman study is the most extensive, and will be reviewed in greatest detail. For both outward WATS and inward WATS, the Feldman study specifies equations for (1) the total number of WATS lines, (2) the number of lines with measured service, (3) the number of lines with full service,[23] (4) the total number of messages, and (5) the number of hours of use per line for lines with measured service. As in the analysis of intrastate toll demand, double-logarithmic demand functions are postulated with WATS price, DDD price, personal income, an exchange rate proxy, and regional dummy variables as independent variables. Intrastate cross-sectional data for forty-one states for November 1973 are used in estimation. The demand functions obtained thus refer to intrastate WATS demand as opposed to interstate.

The price and income elasticities from the Feldman study are tabulated in Table 4-6. In addition to the own-price elasticities, cross elasticities with respect to the price of DDD and full or measured WATS (where relevant) are also presented.[24] While it might seem at first glance that estimated own-price elasticities are implausibly large, it must be kept in mind that businesses subscribe to WATS as

---

[22] This is clearly a simplification, for many business customers subscribe to WATS and private line, in addition to having access to MTS. The relevant question, in this case, is how the optimal mix of MTS, WATS, and private line is determined. As was mentioned in Section 8 of Chapter 2, theoretical analysis of this question is noteworthy for its absence.

[23] Full-service lines refers to lines with unlimited use. With measured WATS, there is an initial allowance of ten or fifteen hours; subsequent hours are then charged for at tenths of hour intervals. Measured WATS is thus like measured local service with an initial allowance.

[24] WATS price is defined as average revenue per (per billing period) line, and DDD price is measured as average revenue per DDD message.

Table 4-6. Estimates of Price and Income Elasticities of Demand for WATS (Feldman Study)

Dependent Variable	*Own-Price Elasticity*	*Cross-Price Elasticity*			*Income Elasticity*
		DDD	*WATS Measured*	*WATS Full*	
Inward WATS					
Total lines	−1.40	1.17	—	—	0.91
Measured lines	−2.34	1.52	—	0.64	0.75
Full lines	−2.83	0.78	1.82	—	1.01
Messages	−0.91	0.69	—	—	1.02
Hours/measured line	—	—	—	—	0.13
Outward WATS					
Total lines	−2.30	2.11	—	—	1.06
Measured lines	−3.25	2.07	—	0.88	0.78
Full lines	−3.37	1.90	1.08	—	1.09
Messages	−1.15	0.96	—	—	0.97
Hours/measured line	—	—	—	—	0.32

Source: Feldman (1976).

a substitute for regular long-distance service. This is true for inward WATS, as well as outward WATS, since the alternative may be the acceptance of a large number of reversed-charge calls. Thus, the estimated price elasticities appearing in the table clearly present a picture of substitution. The large positive cross elasticity with respect to the price of DDD indicates that WATS and regular toll are strong substitutes. Not surprisingly, measured WATS and flat-rate WATS are also indicated to be substitutes, but the substitution is not as strong as for either measured WATS with regular toll.[25] The weakest equations are the ones for hours per measured line. Price is absent from these equations, and income is of little importance. The absence of WATS price is surprising (to me, anyway), and may result from the use of an average price for the price variable, rather than the marginal price, which is theoretically more appropriate. Finally, most of the estimated income elasticities are seen to be in the neighborhood of 1.

Davis et al. (1973) estimate, as part of the AT&T FORECYT model, equations for deflated WATS revenue, deflated private-line revenue, and the number of private-line telephones. Logarithmic Koyck models are specified and estimated with quarterly data over the period 1961 through 1972.[26] In addition to the lagged dependent variable, independent variables are a WATS price index and sales of domestic airlines (since airlines are very heavy users of WATS) in the WATS equation, and a price index for private line and the number of private-line telephones in the equation for deflated private-line revenue.[27] The price and income elasticities estimated

[25] Cf. the cross-price elasticities for DDD and the other WATS offerings in the equations for measured lines and full lines. The own price appears in the equations directly and as a deflator of the price of substitutes. In the equation for measured outward WATS, for example, the price variables are (all as logarithms): the price of measured WATS, the price of DDD divided by the price of measured WATS, and the price of flat-rate WATS divided by the price of measured WATS. The coefficient for the price of measured WATS then measures the own-price elasticity when the prices of measured WATS, flat-rate WATS and DDD all change in the same proportion. This elasticity is quite small (−0.30 in the case considered). Moreover, the *t*-ratio for this coefficient is always smaller than are the *t*-ratios for the price substitution terms. This is further evidence of the importance of substitution among the three forms of long-distance communication.

[26] The WATS revenue data are interstate WATS revenue. The private-line data, in contrast, refer to revenue from all Bell private-line telephones, local, intrastate, and interstate.

[27] The price indices for WATS and private-line revenues are deflated by the implicit deflator for GNP. Presumably, airline sales are also deflated, although the authors do not state that this is the case.

for WATS and private-line revenues are reproduced in Table 4-7.[28]

The WATS elasticities for both price and income agree very well with the elasticities that have been estimated for MTS. This is also the case for the long-run price elasticity for private line (cf., especially, Table 4-1). However, this is what we should expect, since, as we have just been discussing, WATS, private line, and regular MTS serve essentially the same ends. That the short-run price elasticity for private line is quite large in relation to its counterpart for WATS suggests that there is a much faster rate of adjustment in demand to a price change in the private-line market than in the WATS market.

Subissati (1973) estimates equations for deflated WATS revenues (inward and outward combined) and deflated business toll revenue using monthly data for Bell Canada for the period 1967 through 1972. The settlement data used are from Bell Canada to the rest of Canada. Independent variables in the WATS equation are a WATS price index, the value of retail trade in Ontario and Quebec, [29] the two-quarter change in (deflated) business toll revenue, and a time trend. The WATS price index is deflated by a similar index for business toll, and retail trade for Ontario and Quebec is deflated by the national CPI. The time trend is intended to reflect the more rapid rate of growth in inward WATS as compared with outward WATS. Finally, the two-month change in business toll revenue is intended to capture the build-up in business toll calling that is required before the use of WATS becomes cost effective.

Table 4-7. Estimates of Price and Income Elasticities of Demand for WATS and Private Line (Davis et al. Study)

	Price Elasticity		*Income Elasticity*	
Service	*Short-Run*	*Long-Run*	*Short-Run*	*Long-Run*
WATS	−0.17	−0.84	0.18	1.04
Private line	−0.74	−1.03	—	—

Source: Davis et al. (1973, p. 38). The elasticities in the equation for private-line revenues are conditional on the number of private line telephones.

[28] The coefficient for the lagged dependent variable in the equation for private-line telephones is so close to 1 that the calculation of a "long-run" elasticity is not a meaningful exercise. The short-run price elasticity is estimated to be −0.009.

[29] Retail trade is introduced as a seven-month moving average, beginning in t-2 and going backward in time, using an "inverted" V lag structure.

Independent variables in the business toll equation are a business toll price index, manufacturing shipments (for all of Canada), and the first difference in (deflated) WATS revenue. The business toll price index is deflated by the national CPI. Manufacturing shipments are included as a five-month moving average, beginning at t-5 and going backward in time, using an "inverted" V lag structure. The first difference in WATS is intended to reflect the substitution between WATS and regular toll by business.

Because of the presence of WATS revenue in the toll equation, Subissati's model is simultaneous, and two-stage least squares is used in estimation. Subissati finds significant negative price effects in both equations, as well as significant income effects.[30] The coefficient for the two-month change in business toll revenue is also positive, in line with Subissati's a priori reasoning, but, surprisingly, the coefficient for the first difference in WATS revenue in the business toll equation is also positive. Subissati suggests, among other things, that this may be a reflection of a "free-calling" attitude developed by flat-rate billing. Also, since a WATS line serves only one call at a time, frequent queuing may occur, and many calls may be sent by MTS, rather than waiting for their turn on the WATS line.

Despite the illuminating results obtained in the Feldman study as regards the substitution among regular WATS, measured WATS, and regular toll service, our knowledge of the demand for WATS and private-line service is, in general, very unsatisfactory. The biggest unknown is private line. The study of Davis et al. provides some evidence that the long-run price elasticity for real private-line revenues, holding the number of private-line telephones constant, is well within the range that has been estimated for the long-run price elasticity for regular toll. This is essentially all that the existing empirical evidence can tell us about private-line demand. The equation for private-line telephones in the Davis et al. study is very unsatisfactory, but this equation appears to be the state of the art. Neither do we know anything about the cross-price elasticities of demand between private line and WATS and between private line and MTS. And with the strong current interest in restructuring intrastate private-line rates—in some states, there are requests to increase certain private-line rates in the range of 300 percent—this gap in knowledge is strikingly evident.

---

[30] The own-price elasticities estimated by Subissati are −0.33 and −0.25 for WATS and business toll, respectively, where the cross elasticity for WATS with respect to the price of business toll is estimated to be 0.33. The estimated income elasticities are greater than 1 in both equations.

COIN STATIONS

Recent years have seen concerted efforts by telephone companies to increase the charge for local calls dialed from coin stations. However, as noted in Section 1 of Chapter 3, the literature on the demand for coin-station services is small and unsystematic. In my search through the literature, I have found only one study that involves explicit econometric analysis of the demand for coin-station services, this being in the study of Feldman (1976).[31] The focus in the Feldman study is on intrastate toll demand originating from coin stations, and models are estimated for (1) the number of operator-handled calls, (2) the number of person-to-person calls, (3) the average length of conversation for operator-handled calls, and (4) the average length of conversation for person-to-person calls. Intrastate Bell data for the fourth quarter of 1973 for the forty-eight contiguous states in the United States are used in estimation.

In the equations that have the number of calls as the dependent variable, the independent variables are average revenue per call (as a measure of price), average length of haul, population (Bell area), and regional dummy variables. The average revenue per operator-handled call is an additional predictor in the equation for person-to-person calls. Independent variables in the equations for average length of conversation are average revenue per minute, average length-of-haul, per-capita personal income, and the regional dummy variables. Double-logarithmic functions are specified throughout.

The price and income elasticities obtained in the study are tabulated in Table 4–8. The substantial price elasticity in the equation for person-to-person calls is something of an illusion, for it reflects a large cross elasticity with respect to the price of a station call. This cross elasticity is estimated to be 2.04. The "pure" price elasticity is thus -0.82, in keeping with the price elasticity in the equation for station-to-station calls, which includes only the own-price. In neither of the equations for average length of conversation is income very significant statistically. Its *t*-ratio in the equation for person-to-person calls is 1.50 and only 0.20 in the equation for station calls. However, that income is more important in the person-to-person equation clearly makes sense.

It is interesting to compare Feldman's results for toll calls made from coin stations with those for regular intrastate toll (cf. Table 4–3).

---

[31] Brubacher and Hiscock (1978) provide an analysis of the impact of the increase in the price of a local call from 10 cents to 20 cents in 1974 and the subsequent return to 10 cents in 1977 in the parts of Ontario and Quebec served by Bell Canada. Box-Jenkins techniques are used in the analysis. Unfortunately, I came across this study too late to include it in the discussion.

Table 4-8. Estimates of Price and Income Elasticities of Demand for Pay-Station Toll Calls (Feldman Study)

Dependent Variable	*Price Elasticity*	*Cross-Price Elasticity*	*Income Elasticity*
Number of Calls			
Station	−0.88	.04	—
Person-to-person	−2.86	2.04	—
Average Length of Conversation			
Station	−0.19	—	0.03
Person-to-person	−0.78	—	0.31

Source: Feldman (1976).

In general, price elasticities for coin-station toll calls are somewhat larger than for toll calls from noncoin phones, while income elasticities are somewhat smaller. The most marked difference involves the coefficient for the exchange area proxy. In the noncoin toll equations, the coefficient is positive (and highly significant), whereas in the coin-station equations, it is negative (and also highly significant). It is not clear what this might signify.

DEMAND FOR VERTICAL SERVICES

Vertical services encompass all equipment at the subscriber end, except for the main telephone. Included in vertical services, therefore, are residence and business extensions, PBX stations, PBS switchboards and switching equipment, Centrex, residence package plans, and other service and supplementary telephone equipment (hold features, intercom systems, special lights, bells, chimes, whistles, Mickey Mouse wagging his tail, etc.). The econometric literature on the demand for vertical services includes part or all of six studies, Feldman (1976), Huang (1975), Kearns (1978), New York Telephone (1976b), Reitman (1977), and Southern New England Telephone (1977).

The Feldman study specifies equations for business and residence extensions separately, while Huang and Kearns both focus on residence extensions. The Feldman study uses state cross-sectional data in estimation, while Huang uses monthly time-series data for Minnesota. Reitman's primary focus, too, is on residence extensions, but an equation for the residential demand for total vertical services is also estimated. Reitman uses quarterly time-series for forty-five states for the period 1966–1974 in estimation. The NYT study estimates a model for total vertical services (business and residence combined) for New York State for the period 1965:2–1975:2. Finally, the SNET study estimates a model for total vertical services using quarterly data for the period 1970:2–1976:4.

Independent variables in the Feldman (double-logarithmic) equations are revenue per extension (as a measure of the price of an extension), the price of local service, population, per-capita personal income, and regional dummy variables. The estimated elasticities with respect to the price of local service are −0.17 and −0.31 for residence and business, respectively. Kearns estimates a pooled time-series/cross-sectional model, using a data set consisting of observations on forty-seven geographical entities (states and cities) for thirty-seven quarters. The dependent variable in the analysis is deflated residence-extension revenue per capita, while the independent variables are per-capita disposable income, the monthly basic service charge, the price of a residence extension, and the service connection charge (all of the prices are deflated by the CPI). The charge for additional message units (where applicable) was also included as a predictor, but its coefficient was insignificant, and the variable was dropped in the final model. A number of analysis-of-covariance models were used in the pooled analysis, and, in general, the results indicate that demand structures vary across states.

Huang estimates models for total residence extensions, inward movement of residence extensions, and outward movement of residence extensions. (The difference between inward and outward movement equals net change in extensions.) Independent variables are a price index for extensions and a measure (or measures) of the level of economic activity (usually residential construction). The number of residence main stations is also included in the model for total extensions. Static double-logarithmic equations are estimated using monthly time-series data. Autocorrelation is rampant in each case, and a correction for autocorrelation is applied. The strongest results are obtained for inward and outward movement.

The NYT study employs a logarithmic Koyck distributed-lag model, with a price index for terminal equipment, personal income, time, and a dummy variable to adjust for the effects of the fire in 1967 as predictors. The terminal-equipment price index and personal income are both deflated by the CPI for the New York City metropolitan area. The dependent variable is defined as vertical-services revenue deflated by the price index for terminal equipment. Independent variables in Reitman's equations are a price index for residence extensions, per-capita personal income, and the number of residence main stations. The extension price index and personal income are both deflated by the national CPI. The equations are double-logarithmic, and are estimated using quarterly time-series data for each state separately and for all states combined. Reitman's results for all states combined are, in general, much stronger than for the states individually. In the combined equation, it is assumed that structures are homogeneous across states, which, in light of the results obtained by Kearns, is probably not a valid assumption. Finally, the independent variables in the SNET study are real disposable income for Connecticut, a price variable, and the dependent variable lagged one period.

The price and income elasticities obtained in the six studies are tabulated in Table 4-9. The price elasticities are all less than 1 (in absolute value), suggesting that the demand for vertical services is inelastic with respect to changes in price. Kearns obtains an estimate of −0.33 for the price elasticity for residence extensions in his covariance model. Reitman obtains an estimate of −0.45. Kearns also finds a significant cross-price elasticity of −0.29 with respect to the monthly basic service charge. With exception of Feldman's equation for business extensions and Huang's equation for outward movement, [32] the estimated income elasticities are also less than 1. Kearns's estimate of the income elasticity for residence extensions is 0.25.

THE IMPACT OF SERVICE-CONNECTION AND OTHER NONRECURRING CHARGES

The impact of service-connection and other nonrecurring charges on demand is of a great deal of current interest because of accumulating pressure to disaggregate local-service charges and, in particular, to bring installation charges into line with costs. The move toward cost-based installation charges became a serious matter on September 27, 1972, in Colorado when sharply high service-connection charges were put into effect.[33] The charge for installing a telephone increased from a minimum of $7 to a minimum of $23 for a residence main station, while for a business, the minimum installation charge went from $7 to $15 for a phone on a private branch exchange

[32] Huang's income elasticity for outward movement seems implausibly large.

[33] As stated in Decision No. 81320 [Docket 717, Colorado Public Utility Commission (1972, p. 16)] "Failure to charge reasonably in line with the expense incurred is unjustly discriminatory to those that do not require the service, or at least require such services infrequently."

Table 4-9. Estimates of Price and Income Elasticities of Demand for Vertical Services

Study	*Dependent Variable*	*Price Elasticity*	*Income Elasticity*	*Type of Data*
Feldman (1976)	business extensions	−0.41	1.20	CS: states
Feldman (1976)	residence extensions	−0.16	0.52	CS: states
Huang (1975)	residence extensions	−0.008	0.002	TS: M, Minnesota
Huang (1975)	residence extensions (inward movement)	−0.26	0.44	TS: M, Minnesota
Huang (1975)	residence extensions (outward movement)	0.77	2.06	TS: M, Minnesota
Kearns (1978)	deflated residence-extension revenue per capita	−0.33	0.25	TS-CS: Q, 45 states
NYT (1976b)[a]	deflated terminal equipment revenue	(−0.16) −0.59	(0.22) 0.82	TS: Q, New York
Reitman (1977)	residence extensions	−0.45	0.52	TS-CS: Q, 45 states
Reitman (1977)	total residential vertical services	−0.99	0.29	TS-CS: Q, 45 states
SNET (1977)[b]	deflated total vertical services	(−0.07) −0.15	(0.34) 0.79	TS: Q, Connecticut

[a]The NYT model is dynamic; short-run elasticities are in parentheses;

[b]The long-run in the SNET study refers to the elasticity calculated after the lapse of one year.

Symbols: TS: time-series; CS: cross-section; M: monthly; Q: quarterly.

(PBX) and to $60 for a non-PBX phone.[34] The charge for disconnecting a residence phone, which previously was zero, was set at $12.

The Colorado case has attracted considerable attention, and at least one formal study in an analysis by S. K. Black and P. V. Tyron, published in June 1976 by the Office of Telecommunications Policy of the U.S. Department of Commerce. Black and Tryon estimate econometric equations for both inward and outward movement of main stations and extensions for both business and residence customers, the purpose of the analysis being to estimate the impact of the September 1972 increase in service-connection charges.[35] Black and Tryon specify equations for inward and outward movement, with electric-meter installations, time, seasonal dummy variables, and a dummy variable for the change in the service-connection charge as independent variables. Electric-meter installations are included as a proxy for the economic and demographic factors that influence telephone movement. Monthly data for the period January 1969 through September 1973 are used in estimation.

Based on the coefficient for the dummy variable for the increase in the service-connection charge, Black and Tryon estimate that the increase in the service-connection charge resulted in a decrease in inward movement of 4,800 telephones per month and a decrease in outward movement of 2,050 telephones per month. The net effect of the increased service-connection charge is thus estimated to have been 2,750 fewer new telephones per month. Of the decrease of 4,800 in inward movement, Black and Tryon estimate that 1,500 were potential residence main-station customers, 1,700 were potential residence extension customers, 400 were potential business main-station customers, and 1,200 were potential business extension customers.[36]

Unfortunately, Black and Tryon do not list the regression equations that generated these estimates, so that an evaluation of the statistical quality of the results is not possible.[37] Clearly, however,

---

[34] The maximum installation charge, if rewiring is necessary, was set at $42 for a residence phone and $55 for a PBX business phone. For a non-PBX business phone, $60 is the maximum as well as the minimum.

[35] Logically, the Black-Tryon study belongs in Section 2 of Chapter 3 with the analyses of access demand. I have included it in this section because of its explicit focus on service-connection charges.

[36] See Black and Tryon (1976, p. 7).

[37] Black and Tryon provide 95 percent confidence intervals for their forecasts. However, estimates of the other coefficients, standard errors, and so on should have been provided as well.

an increase in service-connection charges of the magnitude of the one in Colorado should be expected to have an impact on inward and outward movement. Yet, at the same time, one should also expect the elasticities to be small.[38] And the price elasticities for main-station inward movement implicit in the Black-Tryon results clearly confirm this, being of the order of -0.02 for residential customers and -0.06 for business customers. These values agree closely with the estimates obtained by Perl (1978) and Waverman (1974).[39]

The only other econometric analysis of non-recurring charges that has come to my attention is a study undertaken by the Southern New England Telephone Company (1977a). The dependent variable in the SNET analysis is deflated nonrecurring revenues in the area served by SNET in Connecticut. Independent variables are a deflated index of nonrecurring charges, disposable personal income, the change in Connecticut unemployment, the dependent variable lagged one period, and two dummy variables to correct for strikes in 1968:2 and 1971:3. A double logarithmic model is estimated with quarterly data for the period 1963:2 through 1976:4. The estimate of the price elasticity in the SNET model is -0.06 in the short run and -0.10 after the lapse of one year. Although they are larger, these values are, in general, in keeping with the values obtained by Black and Tryon, Perl, and Waverman. The one-year income elasticity is estimated to be 0.72. The *t*-ratios for the price and income coefficients are -2.72 and 2.53, respectively.

INTERNATIONAL DEMAND

In previous sections, a number of studies of telephone demand in other countries have been reviewed, but the focus in each case was on domestic demand. Our attention in this section will be on international demand, where international is defined as communication to any point outside of the forty-eight contiguous states. As was noted in Chapter 1, the literature on international demand is modest in comparison with the literature on domestic demand. One could hardly expect otherwise. However, between 1950 and 1975, international messages from the United States increased an average

---

[38] As was discussed in Section 2 of Chapter 3, we should expect "churning" caused by temporary changes of address (college students leaving for the summer, use of vacation homes, etc.) to be the most sensitive to increases in service-connection charges. Inward and outward movement associated with permanent changes of address will be less sensitive to changes in these charges because they are part of the larger costs of moving. Also, one should expect changes in service-connection charges to have a greater impact on extension phones than on main stations.

[39] See Section 2 of Chapter 3.

18.6 percent per year. By comparison, domestic U.S. toll traffic in the over-2,000 mileage band (the fastest-growing domestic band) increased an average of 15.1 percent per year. In 1975, gross revenue from international MTS messages was 21 percent of gross revenue from domestic interstate-MTS messages. If present trends should continue, international revenues would overtake interstate revenues sometime between 1995 and 2000. Thus, in view of the continuing rapid rate of growth of the international market, the study of international demand is certain also to continue to grow.

Let us begin with some general observations. Since an international telephone call is long-distance by definition, international demand should be expected to behave somewhat the same as domestic toll demand. However, depending on where one is calling, there are also some important differences. In the first place, domestic toll calling is routine with nearly everyone, but this is not necessarily the case with international calling; in fact, most people have never made an international call.[40] In the second place, even when adjusted for distance, an international call is about two to three times more expensive than a domestic toll call, and this may contribute to the mystique that makes an international call seem to be a very different product from a domestic toll call.[41] Also, with the exception of calls to Canada, an international call is typically a production compared with a domestic call. An operator is often required in the country originating the call, and, partly because of time differences, chances of reaching the party desired are usually substantially reduced. For example, the completion rate for domestic calls is about 70 percent, but usually around 50 percent for international calls. Moreover, for calls to some international points, equipment is limited, and it may be hours (or, in some extreme cases, even days) before a call can be placed. Thus, the time required to initiate and complete an international call is an important addition to the out-of-pocket costs of the calls.[42]

---

[40] This excludes calls between the United States and Canada. Also, the focus here is on calls made by U.S. residents. The statement may therefore not apply to residents of Western Europe.

[41] I have in mind the higher priority and the greater sense of urgency usually associated with an international call, particularly by customers who do not frequently use this service. A similar mystique attaches, although less now than in the past, to a long-distance call vis-à-vis a local call. In the past, the phrase, "long-distance calling", was virtually certain to snap one to attention. With DDD, the stimulus has become more subtle; one listens for the long-distance "hum." For a discussion and analysis of a survey conducted in 1955 of motives and attitudes toward long-distance calls, see Adelson et al. (1956).

[42] However, it should be noted that, with the advent of modern cable tech-

Another substantive difference between international and domestic toll demand is the possible importance of telegraph and telex as competitive modes of communication. Much of the time, the purpose of an international call may be served equally well by a telegram or telex. This contrasts with domestic toll calls, where the time and out-of-pocket cost of a call are small in comparison with the costs of a telex or telegram. Recognition of the competition provided by telegraph and telex has led, in some cases, to the specification of multi-equation models in which the international demands for telephone, telegraph, and telex are approached as a group.[43]

Much of the research on international telephone demand has been undertaken at AT&T Long Lines and Bell Canada,[44] so that, not surprisingly, most of the focus in the empirical literature is on the United States and Canada. In general, research at Long Lines has focused on the development of models, on a country-by-country basis, for explaining the volume of voice communication between the United States and other countries. As of December 1977, Long Lines had developed and were using some thirty-one different models for forecasting traffic between the forty-eight contiguous states and the rest of the world. Of these, twenty-five are annual models, three quarterly, and three monthly. Included in the "inventory" are models for twenty-eight countries, Alaska, Hawaii, Puerto Rico, and the Virgin Islands.

Besides the efforts of Craver and Khadem, other studies of international telephone demand include Communications Satellite Corporation (1969), Kumins (1967), Rea and Lage (1978), Lago (1970), Naleszkiewicz (1970), and Yatrakis (1972). As mentioned earlier, the studies of Rea and Lage, Lago, and Yatrakis focus on telegraph and telex, as well as on telephone. In each study, there is an attempt to estimate cross-elasticities of telephone demand with respect to the prices of telegraph and telex. Telegraph and telex as substitutes for voice communication are not considered in the models

---

nology (TAT III in 1963) and satellite communications in 1965, both the completion time and the real cost of an international call have been significantly reduced. Even the proportion of calls dialed by the customer has increased sharply in the last few years. In 1973, for example, only 35 percent of international (two-way) calls (not including Canada and Mexico) did not require an operator in either country. By 1976, this figure reached 60 percent. I am indebted to Robert Craver of the International Department, AT&T Long Lines for these observations.

[43] See, in particular, Rea and Lage (1978), Lago (1970), and Yatrakis (1972).

[44] See Craver 1975, 1976a, 1976b, 1976c, 1976d, 1977) and Khadem (1973b, 1976).

of Craver and Khadem.[45] The studies of Rea and Lage, Lago, and Yatrakis also differ in the data that are used in estimation. In the studies of Craver and Khadem, the models are either of one country with another or of one country with the rest of the world, and time-series data are used in estimation. In the Rea-Lage, Lao, and Yatrakis studies, in contrast, the focus is not on specific country pairs, and either cross-section or pooled time-series/cross-section data are used in estimation.[46]

The price and income elasticities estimated by Craver, Khadem, Rea and Lage, Lago, and Yatrakis are tabulated in Tables 4-10 and 4-11. The estimated elasticities of Craver and Khadem are given in Table 4-10, while the estimates of Rea and Lage, Lago, and Yatrakis are given in Table 4-11. The results suggest price and income elasticities for international telephone demand that, on balance, are larger than the ones for domestic toll demand.[47] This is especially evident in Khadem's results for Canada. A second point to note is the impact of the prices of telegraph and telex. Both Lago and Yatrakis find a positive cross elasticity with respect to the price of telegraph, but Rea and Lage find this cross elasticity to be negative.

---

[45] AT&T Long Lines feels that competition between international calls and telegraph or telex is not a significant factor. The use of international telegrams has been declining for several years, and is of little quantitative importance today. And telex services require special terminal equipment and are thus used almost exclusively by business. Indeed, in AT&T's view, the business use of telex services is more a complement to voice communications, rather than a substitute for, because of a customer need to have a paper copy. Again, I am indebted to Robert Craver of AT&T Long Lines for these observations.

[46] Rea and Lage use a pooled time-series/cross-sectional sample of observations for thirty-seven major routes. The data are annual for the years 1964-1973. Lago also uses a pooled time-series/cross-sectional sample consisting of annual observations on telephone traffic between the United States and twenty-three other countries for 1962-1964. Finally, Yatrakis uses a cross-sectional sample consisting of observations for sixty-six countries for 1967. The dependent variable for Yatrakis is the number of originating telephone calls from a country to all other countries to which it is linked via good-quality telephone facilities.

[47] There appears to be a major anomaly between the estimates of the income elasticities in Craver's equation for billed minutes (U.S. to U.K.) and his equation for the number of messages (U.S. to U.K.). The income elasticity is 0.09 in the former and 2.67 in the latter. In the first equation, income is measured by the volume of trade between the United States and the United Kingdom, and a time trend is also included, which is the dominant predictor. In the second equation, income is measured by U.S. GNP, and there is no time trend in the equation. Both equations include variables that account for circuit constraints. The second equation—that is, the one with the larger income elasticity—is definitely the superior specification.

Table 4-10. Estimates of Price and Income Elasticities for International Telephone Demand (Part I)

Study	Dependent Variable	Price Elasticity		Income Elasticity		Type of Data
		Short-Run	*Long-Run*	*Short-Run*	*Long-Run*	
Craver (1976c)	billed minutes, U.S. to U.K.	(−0.38)		(0.09)		TS: A
Craver (1977)	messages, U.S. originating	(−0.51)		(2.37)		TS: A
Craver (1976d)	messages,	(−0.46)		(2.67)		TS: A
Craver (1976d)	messages, U.K. to U.S.	−0.63	−1.02	—	—	TS: A
Craver (1976a)	messages, 2-way U.S. to Canada	−0.23	−0.62	0.44	1.18	TS: A
Craver (1976a)	messages, 2-way U.S. to Canada	−0.42	−0.52	1.28	1.60	TS: Q
Craver (1976a)	messages, U.S. to Canada	(−0.70)		(1.65)		TS: Q
Craver (1976a)	messages, Canada to U.S.	(−0.55)		(1.49)		TS: Q
Khadem (1976)	messages, Canada outward	−0.89	−3.65	2.19	9.03	TS: Q
Khadem (1973)	messages, Canada to U.S.	—	—	0.59	1.84	TS: Q
Khadem (1973)	deflated ARPM Canada to U.S.	−0.61	−1.67	—	—	TS: Q
Khadem (1973)	deflated revenues Canada to U.S.	−0.30	−1.12	0.94	3.42	TS: Q

Symbols: TS: time-series; A: annual; Q: quarterly; ARPM: average revenue per messages.

Table 4-11. Estimates of Price and Income Elasticities for International Telephone Demand

Study	*Dependent Variable*	*Telephone*	*Telegraph*	*Telex*	*Income Elasticity*	*Type of Data*
Rea & Lage (1978)	messages, U.S. outward	−0.73	−2.93	0.31	2.62	CS-TS: A, 37 point-to-point routes
Lago (1970)	messages, U.S. outward	−1.69	0.29	—	—	CS-TS: A, 23 countries
Yatrakis (1972)	messages, outward	−0.82	1.31	−0.49	0.66	CS: 46 countries, 1967
Yatrakis (1972)	messages, outward	−2.31	2.73	−0.42	1.78	CS: 23 developed countries, 1967
Yatrakis (1972)	messages, outward	−0.96	—	—	—	CS: 23 developing countries, 1967

Symbols: TS: time series; CS: cross-section; A: annual.

Of the two positive estimates, the one obtained by Yatrakis is the only one with robustness, for the *t*-ratio for Lago's cross elasticity is only 1.7. On the other hand, the cross elasticity estimated by Rea and Lage is not only negative, but is also very large (−2.93) and highly significant statistically (its *t*-ratio is −7.52). However, in general, one has to view this finding with skepticism.

Despite the attention that international telephone demand has received in recent years, much remains to be done. Existing results clearly indicate that price—as well as income—is an important factor, and, further, that the prices of telex service should also be considered in order to better determine whether telephone and telex are substitutes or complements. The results tabulated in Tables 4–10 and 4–11 suggest that international price and income elasticities—especially income elasticities—may be larger than their domestic counterparts. Given that an international call is clearly a luxury in most quarters, especially residence calls, this is not unreasonable. Nevertheless, one must look with skepticism at this point at the size of the elasticities that have been obtained. The problem, as has been noted by Kumins (1967), is that most rate changes to date have been rate decreases associated with the introduction of new or expanded facilities. Consequently, it is not clear how much of the elasticity that is observed is the result of the releasing of a supply constraint, as opposed to a genuine demand response. While there has been some attempt to take circuit constraints into account [notably by Craver (1976d) and Kumins (1967)], the problem clearly requires more attention than it has been given to date.

7. DEMAND FOR YELLOW-PAGES ADVERTISING

Up to this point in this review, we have considered the demand for access to the telephone system, the demand for the use of the telephone system, and the demand for certain types of telephone-related equipment. However, there is another service that telephone companies provide, but which is usually not considered in the same light as access and use, namely, yellow-pages advertising. Yellow-pages advertising is an important source of revenue for telephone companies. This section, accordingly, is devoted to a brief discussion of a recent study by Wang (1976) of the demand for yellow-pages advertising in the service areas of fourteen Bell System operating telephone companies.

Since the topic is somewhat off the beaten path, I shall make the digression self-contained by considering a theoretical framework for analyzing the demand for Yellow-pages advertising. Accordingly, let us assume (following Wang) that the demand for yellow-pages advertising can be reduced to the demand for yellow-pages space and that the price charged for an advertisement is proportional to the space the advertisement requires. Space, however, is not perfectly divisible, but rather is offered for sale in blocks of different sizes, say, $s_1, s_2, \ldots, s_n$. Let p_i be the price of s_i, and assume that the s_i are ordered $s_1 < s_2 < \ldots < s_n$, so that $p_1 < p_2 < \ldots < p_n$.

Let us now assume that we have a population consisting of M firms and that each firm allocates a maximum amount that it is willing to spend for space in the yellow pages.[48] Let this amount for the jth firm be denoted by y_j. On the assumption that more space is preferred to less space, the firm will purchase the largest advertisement that it can afford. That is, the firm will purchase the maximum s_i for which $y_j \geqslant p_i$. Alternatively, s_i will be determined as the i for which

$$\begin{aligned} &p_i \leqslant y_j < p_{i+1} && \text{for } i \leqslant n-1 \\ &p_i < y_j && \text{for } i = n. \end{aligned} \tag{49}$$

The number of advertisements of size s_i that will be purchased will then be given by the number of firms whose yellow-pages advertising budgets satisfy these inequalities.

More formally, let

$$\delta(p_i, y_j) = \begin{cases} 1 \text{ if } p_i \leqslant y_j < p_{i+1} \text{ or } p_n < y_j \\ 0 \text{ otherwise,} \end{cases} \tag{50}$$

and let

$$N_i = \sum_{j \in M} \delta(p_i, y_j). \tag{51}$$

Further, let $F(y_j)$ denote the cumulative distribution function of yellow-pages advertising budgets over the M firms. Then,

$$N_i = F(p_{i+1}) - F(p_i) \tag{52}$$

defines thee demand function for the ith size of yellow-pages advertisement.[49]

[48] What determines this maximum amount is clearly an interesting and relevant question, but will not be considered here.

[49] The yellow-pages space of size s_i demanded will clearly be given by $N_i s_i$.

If we now assume that the distribution of yellow-pages advertising budgets is such that

$$F(p_{i+2}) - F(p_{i+1}) < F(p_{i+1}) - F(p_i) \tag{53}$$

for all i, then $N_i > N_{i+1}$, and N_i plotted against p_i will be a decreasing step function.[50] If the size of an advertisement were made to vary continuously, and if the number of firms is large, this step function would approach a smooth curve. This curve, which we shall denote by $N(p)$, can be viewed as the market demand function for yellow-pages advertising.

Let me now turn to the empirical results obtained by Wang. Wang assumes that the function $N(p)$ can be approximated by a double logarithmic function,

$$lnN = \alpha + \beta lnp + \gamma lnT + u, \tag{54}$$

where T and u denote the number of telephones and a random error term. The number telephones is included as a surrogate for market size. Wang estimates the model in (54), using pooled time-series/cross-sectional data[51] for fourteen Bell System operating telephone companies.

Wang's price elasticities vary from -0.72 to -1.49, and his elasticities with respect to the number of telephones range from 0.92 to 1.26. The medians of these estimates are -0.94 and 1.02. Wang's results will be illustrated with his equation for one OTC, namely:[52]

$$\begin{aligned} ln\hat{N} &= -4.33 - 0.72 lnP + 1.14 lnT. \\ &\quad (-17.81)(-39.13) \quad (52.08) \end{aligned} \tag{55}$$

Wang does not present any goodness-of-fit statistics, but the large values for the t-ratios suggest that the equation probably fits the data very well. This is the case in all of the equations estimated. The coefficient of 1.14 for the logarithm of the number of telephones is of particular interest, for it is significantly greater than 1, and is therefore consistent with the existence of a positive network externality.

---

[50] A sufficient condition for this result is that the "yield" from a yellow-pages advertisement be a decreasing function of the ad's space.

[51] The dependent variable in the regression equations is the number of advertisements of size (i.e., price) i in directory k for year t. The number of telephones does not vary with i, but does vary with k and t. Price varies with i, k, and t.

[52] T-ratios are in parentheses. This equation is estimated with pooled data from nineteen directories for the years 1970–1976. The sample contains 819 data points in all.

8. NONECONOMETRIC APPROACHES TO FORECASTING TELEPHONE DEMAND

In the preceding ten sections, we have examined the econometric literature on telephone demand. By "econometric" I mean studies that focus on the quantification of the basic economic, sociological, and demographical determinants of telephone. In most of the studies reviewed, the emphasis has been on estimation of price and income elasticities of demand and the development of models for forecasting the impact on revenues of changes in telephone tariffs. However, when we consider forecasting per se, an econometric model is only one of several approaches that might be used. Time-series models have been used in a number of contexts, and, in this concluding section, we shall provide a brief summary of several of these.[53]

Bhattacharya (1974)

Bhattacharya (1974) is concerned with forecasting the demand for telephones in Australia. To this end, a Box-Jenkins ARIMA[54] model is applied to a demand series from which the effect of the connection charge plus annual rental had been removed. The model estimated thus takes the form of a regression model with one independent variable and an ARIMA error term. The dependent variable in the model is the twelve-month percentage change in the number of new telephones installed, while the independent variable is the twelve-month percentage change in the connection charge plus annual rental. Monthly data for the period July 1962 through June 1971 are used in estimation. The price elasticity estimated by Bhattacharya is −0.91 with a *t*-ratio of −8.27.

Dunn, Williams, and Spivey (1971)

Dunn, Williams, and Spivey (1971) focus on forecasting inward and outward movement of business and residence main stations in three Michigan cities, Battle Creek, Grand Rapids, and Flint. The data analyzed are monthly for the fifteen-year period 1954–1968. A number of forecasting models are examined, including a simple trend-seasonal-irregular time-series decomposition, adaptive exponential smoothing, and adaptive exponential smoothing with exogenous

---

[53] Even more than in preceding sections, the discussion in this section is neither detailed nor exhaustive. The purpose here is merely to provide readers with an introduction to the time-series literature on forecasting telecommunications demand.

[54] Auto*R*egressive *I*ntegrated *M*oving *A*verage. For a discussion of Box-Jenkins procedures, see Box and Jenkins (1976) or Granger and Newbold (1977).

variables.[55] The authors emphasize the problems of obtaining meaningful exogenous variables for local areas, but note the gain in precision that is likely to be achieved if exogenous predictors can be found. In the case of Flint, a 15 percent reduction in the mean absolute forecast error was obtained through the use of the number of employees covered by unemployment compensation and the number of households as exogenous predictors.

Thompson and Tiao (1971)

Thompson and Tiao (1971) bring Box-Jenkins methodology to bear on forecasting monthly inward and outward movements in Wisconsin using data of the Wisconsin Telephone Company for the period January 1951 through October 1966. In contrast to the variables used by Dunn, Williams, and Spivey, inward and outward movements are not decomposed to business and residence customers. Also, the series analyzed refer to the inward and outward movement of all telephones, rather than just main stations. As one might expect, Thompson and Tiao find strong seasonal effects in both inward and outward movement, as well as some effects between months. They test the assumption of constant seasonal effects over time, but find no evidence to refute it.[56]

Tomasek (1972)

Tomasek (1972) is concerned with developing a model to forecast inward movement of main stations in metropolitan Montreal. Tomasek notes that traditional predictors, such as the change in personal income and new residential construction expenditures, are difficult to find for an area as small as the one considered, and thus turns to an autoregressive time-series model. Ten years of monthly data (1961-1970) are used in estimation. Tomasek estimates a regression of inward movement on itself at one, three, five, twelve, thirteen, fifteen, and seventeen months in the past, and he also estimates a Box-Jankins ARIMA model. However, on balance, the forecasts from the ARIMA model are not quite as good as the ones from the pure autoregressive model.

Wellenius (1970)

Wellenius (1970) focuses on the development of a model for forecasting the residential demand for telephones in Santiago, Chile.

[55] For discussion of adaptive forecasting models, see Nerlove and Wage (1964) and Theil and Wage (1964). Cf. also Chen and Winters (1966) and Crane and Crotty (1967).

[56] Thompson and Tiao note that inward and outward movements are probably related—most disconnections lead to later reconnections—but the possibilities of this are not examined.

Chapter 5

Evaluation and Critique

In the two preceding chapters, we have taken an intensive look at the empirical literature on telephone demand. A large number of empirical studies have been reviewed, and numerous estimates of price and income elasticities of demand have been tabulated. This chapter has the task of tying things together, of assessing the state of existing research and suggesting a program for future research. Section 1 of this chapter will be devoted to a critique of the models and methodologies that have been used in analyzing telephone demand. The theoretical considerations discussed in Chapter 2 will provide the framework for this critique. In Section 2 the estimates of price and income elasticities of demand will be evaluated, and some of the more important implications of these results will be discussed. In Section 3 the most pressing problems concerning methodology, data, and research topics will be identified and discussed. Finally, Section 4 will address the question, where should we go from here?

A CRITIQUE OF EXISTING MODELS AND METHODOLOGIES

In Chapter 2 we discussed the attributes of the telephone system that set telephone demand apart from the demand for most goods and services. These, to recall, include (a) the distinction between access to the telephone network and the use of the network once access has been obtained, (b) the existence of access and use externalities, and (c) option demand as a possibly important determinant of access demand. The distinction between access and use is critical, and in my opinion should provide the main point of departure in building models of telephone demand. In practice, this means that the analysis of telephone demand should be approached in stages, with the first stage focusing on the demand for access. The most meaningful quantity to explain in this regard is the number of main-station telephones for residential subscribers and the number of main stations plus the number of PBX extensions for business subscribers.

Stage two of the analysis should then focus on the demand for use. However, the demand for use may itself need to be approached in stages, depending upon how use is priced. On the one hand, if a call is priced on a two-part tariff, in which the price for the initial period differs from the price of an overtime period (as is the case for toll calls in the United States and Canada), then the demand for use should consist of two equations, one that explains the number of calls and one that explains average duration. On the other hand, if a call is not priced on a two-part tariff (as is the case in Sweden and the United Kingdom, for example), a single equation that explains the number of conversation-minutes will suffice.

The access/use distinction is found in studies throughout the literature, but it is center stage in only three, namely, Alleman (1977), Pousette (1976), and Waverman (1974). Alleman restricts his analysis to the demand for basic service (i.e., access), but Pousette and Waverman estimate complete models, which is to say that they estimate equations for use, as well as for access. Unfortunately, the other studies that estimate equations for the number of telephones contain little by way of theoretical motivation. In most cases, the analysis is guided by the general principles of demand theory, but the access/use distinction as a peculiar feature of telephone demand is largely ignored. How system size is tied to use, and how use, in turn, depends upon system size is in general not considered.

Let me illustrate the importance of the access/use distinction with Waverman's model for Sweden. Waverman, to recall, estimates equations for the number of main stations, local use, and toll use. The equations for main stations and local use will ustrate the point that I wish to make. Income is a predictor in both of the equations, the price of local use is a predictor in the equation for local use, but not in the equation for main stations, and the number of main stations is a predictor in the equation for local use. Thus, local use depends both directly and indirectly on income (since local use depends on the number of main stations, which, in turn, depends on income), but only directly on the price of local use. (There is no indirect price effect via the number of main stations in Waverman's model, since main stations do not depend on the price of local use.) Suppose, now, that there is a change in income. What will be the impact on local use as measured (say) by the income elasticity of demand? Waverman's equation for local use is a Koyck logarithmic distributed-lag model, so that three income elasticities can be adduced, a "short-run" short-run elasticity and a "long-run" short-run elasticity, both of which are conditional on the number of main stations, and a long-run elasticity in which the number of main stations is allowed to vary in response to the change in income. Waverman's estimates of these three elasticities are 0.23, 0.32, and 1.25. Thus we see that in Waverman's model the indirect effect on local use of a change in income that arises through the equation for main stations is substantial.

The moral of the story is as follows: If the focus is entirely on estimating the effects on usage of changes in the price of usage, an access/use framework is not critical—so long, that is, as usage is assumed to be conditional on the number of main stations—since any impact of the price change on the number of main stations is probably small enough that it can be ignored. However, if the focus is on estimating the impact on usage of changes in income, an access/use framework *is* critical, because to ignore the indirect effect on usage that arises through the adjustment in main stations is likely to lead, in most cases, to a serious underestimate of the total effect of a change in income.[1]

Let me now turn to the access and call externalities. As was discussed in Chapter 2, the access externality can, in principle, be taken into account by making the demand for access depend on system size. One way that this can be done is to relate the current number of main stations to the number of main stations in the preceding period. If there are no other dynamics, a positive access externality will be reflected in a coefficient on the preceding period's main stations that is greater than 1. The call externality is more problematic to measure, but, as suggested in Chapter 2, it too may be represented by system size.

The evidence regarding the access and use externalities is thus to

[1] The state intrastate toll demand models come to mind at this point. Most of the state models assume toll demand to be dependent on the number of main stations (see Table 4-1 of Chapter 4), which is of course as it should be, but the models do not, in turn, include equations for the number of main stations. As noted in the text, there is no problem so long as the focus is on price changes and there exist good exogenous forecasts of the number of main stations. However, there will probably be occasions when the models will be used to forecast the impact of changes in income, and here care must be taken to allow for feedbacks on the number of main stations. The explicit use of an access/use framework automatically reminds the model builder to do this.

be found in the equations that include a measure of system size as a predictor. This occurs in a number of the state intrastate toll demand models and in the models of Feldman (1976), Davis et al. (1973), Pousette (1976), and Waverman (1974). In most of the models, system size is measured by the number of telephones less residence extensions, but in a few of the state models, the number of households or the population is used as a surrogate. Most of the equations in question refer to the demand for use; the only ones that involve the demand for access are the equation of Davis et al. for the total number of telephones (less residence extensions) and Waverman's equation for the number of main stations in Canada.

In general, however, the evidence concerning the two consumption externalities is inconclusive. The strongest suggestion that the externalities may be of some importance is given in Waverman's equation for local use in Sweden.[2] However, the fact that the existing evidence is weak and mixed is hardly surprising since the empirical literature has not explicitly focused on the consumption externalities as factors to be taken into account. Many of the state toll demand models, for example, assume them away a priori by defining the dependent variable as the number of messages per main station (or price-deflated revenues per main station) and then not including main stations as a predictor. Moreover, it is not the case that the externalities have been considered and then dismissed as being unimportant; in general, they have simply been ignored. Waverman, for example, does not see in his results for Sweden the suggestion that the externalities may be of importance, and Pousette, in his otherwise admirable study, ignores them from the start.

I shall now move on to option demand. The literature does not provide any empirical evidence, even inadvertently, regarding its importance. However, this, too, is hardly surprising, for while option demand is an appealing concept, it does not easily lend itself to measurement. But even so, this is not a reason for it being ignored altogether. In Section 4 of Chapter 2 it was suggested that option demand should be relatively more important in rural exchanges than in urban exchanges, and if this is so, the access-demand elasticity with respect to the price of access should be smaller in rural es-

[2] See Equation (9) in Chapter 3. Additional evidence concerning the externalities is found in the studies of Infosino (1976) and Wang (1976). Infosino finds that the number of local calls per line in a sample of residential customers in Los Angeles and San Francisco is positively related to the telephone density of the exchange, while Wang finds that the demand elasticity for yellow-page advertisements of a given size with respect to system size is greater than 1. This result implies that, with the size of an advertisement and price held constant, the demand for the space is stronger in a larger system.

changes than in urban exchanges. This might be tested with the data set used by Perl. Also, as was mentioned in Chapter 2, it seems to me that option demand may be a factor in many subscribers' apparent preference for flat-rate pricing of local service over measured service. While, it is not clear how this idea might be tested empirically, it has important implications for the pricing of access and is therefore worthy of some attention.

Let me now turn to other matters. A major deficiency in the empirical literature is the treatment of prices. Most telephone services are priced on some form of multi-part tariff, and as discussed in Section 5 of Chapter 2, this has a number of important implications. With a multi-part tariff, one has to distinguish between the marginal price and the intramarginal prices, and, in some cases, the multi-part tariff changes the basic logic of the demand model. A toll call in the United States and Canada, for example, is priced on a two-part tariff, since the price per minute is less for overtime periods than for the initial period. Assuming that the research goal is to explain the number of conversation-minutes, the appropriate procedure (as discussed earlier in this section) is to decompose the number of conversation-minutes into the number of calls and the average duration of a call. The number of calls will then depend (besides on income, etc.) primarily on the price of the initial period, while the average duration of a call will depend primarily on the price of an overtime period.

The principles that underly these conclusions have been discussed in Section 5 of Chapter 2, but it is useful to summarize them again here. When a good is priced on a multi-part tariff, the different parts of the tariff affect a consumer's behavior in different ways. In equilibrium, the consumer equates marginal rates of substitution between pairs of goods to the ratios of their respective *marginal* prices. However, when the good involved is defined in several dimensions, then a tariff that is marginal to the decision in one dimension may be intramarginal or extramarginal to the decision in another dimension. With a toll call, there are two decisions to be made: whether to make the call, and how long to talk if the call is made. The price of the initial period is the price that is most relevant to the decision of whether or not to make the call, the time of day to make the call, and whether or not to direct dial, while the price of an overtime period is the price that is most relevant to the decision of how long to speak.

In general, these considerations are not reflected in the empirical literature. There are only two studies that seem genuinely cognizant of the problems that multi-part tariffs pose, namely, Deschamps

(1974) and Pousette (1976). Deschamps discusses explicitly the complications created by multi-part tariffs, and his study stands alone in this regard. Pousette's actions, on the other hand, speak louder than his words, for although he does not mention the need to distinguish between the price of access and the price of use, his equations for new connections contain a price index that is a weighted average of the subscription fee and the call charge, while his equations for use contain only the call charge. Combining the access and use charges into a single index is not the ideal procedure, but it is certainly a step in the right direction.[3]

The failure to distinguish among the various components of a multi-part tariff usually leads to the use of an average price, and the worst situation is when the average price is obtained by dividing revenues by the quantity that is being explained. An average price for toll calls, for example, is frequently obtained by dividing toll revenues by the number of toll calls. Such an ex post procedure is to be avoided because it necessarily establishes a negative relationship between quantity demanded and price. Perhaps the most serious lapse in this connection is by Beauvais (1977) in his study of the demand for local calls. Beauvais defines an average price by dividing the monthly service charge by the number of local calls. However, as was noted, this creates a potentially very serious bias in the estimate of the price elasticity because most of the variation in the price variable is caused by the variation across subscribers in the number of local calls. Other studies in whic price is calculated as an ex post average price include Feldman (1976), Kwok, Lee, and Pearce (1975), Larsen and McCleary (1970), and Rash (1972).

The state toll demand models whose elasticities are tabulated in Table 4-1 of Chapter 4 all use a price index for intrastate MTS calls for the price variable. Laspeyres indices are used in about two-thirds of the models, and chain-weighted indices are used in the remainder. Never do the weights depend on current-period quantities demanded, so that the price variables used in these models do not suffer from the bias problem just referred to. However, they do have a problem in that the initial-period and overtime-period charges are

[3]The studies of Waverman (1974), Perl (1978), and Larsen and McCleary (1970) should also be mentioned in this regard. Waverman distinguishes between the price of access and the price of use in his equations for Sweden, but not in his equations for Canada and the United Kingdom, and Perl, in his analysis of access demand, treats the service-connection charge separately from the monthly service charge and also distinguishes between exchanges with measured local service and exchanges with flat-rate service. Finally, Larsen and McCleary, in their analysis of toll traffic between pairs of states, examine both the average charge per call and the average overtime charge per call. Unfortunately, though, Larsen and McCleary calculate both measures of price from ex post data.

combined in a single index. This is appropriate for the models in which the dependent variable is price-deflated revenues, in which case the dependent variable is (in principle) a measure of conversation-minutes, but it is not appropriate for the models in which the dependent variable is the number of messages. In this case, the price of the initial period should be separated from the price of an over-time period.

Another shortcoming in the empirical literature is a focus on the number of calls, as opposed to conversation-minutes. As has been noted, when the price of a toll call consists of an initial charge plus a charge that depends on the number of overtime periods, one must explain duration as well as the number of messages. If the price of a call were independent of duration, duration could be ignored, but in most instances, the price of a call does vary with duration, so that the possible dependence of duration on price must be taken into account. Only four studies in the literature do this: Feldman (1976), Gale (1974), Pousette (1976), and Waverman (1974). The studies of Feldman and Gale provide the most detailed analyses and, in both cases, duration is found to be negatively related to price.[4] Pousette and Waverman both explain the number of "pulses" (which is a physical measure of holding time), and duration per se is not singled out for analysis. However, Waverman also estimates an equation for the number of messages, and since the price elasticity in this equation is smaller than in the pulse equation, one can infer that there is also a nonzero price elasticity for duration.

The ignoring of duration is especially apparent in the state toll demand models.[5] None of the state models has an equation that focuses directly on duration; fifteen of the models whose elasticities are tabulated in Table 4-1 of Chapter 4 have the number of messages as the dependent variable, while sixteen have price-deflated revenues (PDR) as the dependent variable. Since toll revenues can be decomposed into the product of the number of calls and the average revenue from a call, price-deflated revenues correspond, in principle, to the product of the number of calls and the average duration of a call. The impact of a price change on average duration is thus reflected in the price elasticities that are estimated in the PDR equations.

Whether the dependent variable in the state models should be the number of messages or price-deflated revenues is a subject of some

---

[4] Gale's study is unique in the literature in that it is the only one that focuses directly and exclusively on the dependence of duration on price.

[5] By state models, I mean the intrastate toll demand models that are tabulated in Table 4-1 of Chapter 4.

debate, and this is an appropriate place to offer a few comments. The state models have been developed mainly as planning tools and, as mentioned in Chapter 4, to isolate the market reactions to tariff changes in rate filings before state Public Utility Commissions. In most cases, their primary use in rate filings has been to estimate the impact on toll revenues of a nonzero price elasticity of demand. For the models with the number of messages as the dependent variables, revenues (after repression)[6] are calculated by multiplying the estimated number of messages by a repriced average revenue per message. For the models with price-deflated revenues as the dependent variable, after-repression revenues are calculated by multiplying the estimate of repression-adjusted real revenue by a price index which reflects the new rates.

In my opinion, neither of these procedures is ideal. The problem is that the number of messages and the average duration of a message respond in different ways to changes in the tariff structure. In the models with price-deflated revenues as the dependent variable, the price of a call is represented by a price index which combines the charge for an initial period with the charge for an overtime period, which means that the responses of messages and average duration are reflected in a single elasticity. In situations where the structure of the tariff schedule is to be changed (the charge for the initial period increased relative to the charge for an overtime period, for example), not allowing messages and duration to respond differently could lead to serious forecasting errors. However, the problem in the PDR models is not so much in the use of price-deflated revenues as the dependent variable, but rather in the use of a single price index. The use of two price indices, one for the initial-period charge and the other for the overtime-period charge, would largely overcome the problem.

The problem in the models with the number of messages as the dependent variable is that an equation is missing, namely, an equation for average duration. Either the models make no allowance at all for the effect of a rate change on duration or else it is assumed that the price elasticity for messages applies also to duration. Neither of these procedures is satisfactory. However, the solution (at least in principle) is readily apparent: estimate an equation for average duration.

In Chapter 2 I mentioned the need to model telephone demand in a dynamic framework. The models using the access/use distinction

---

[6] Repression in this context refers to the impact on revenues of a rate change when there is a nonzero price elasticity of demand.

are dynamic by definition—since use is predicated on system size—and distributed-lag models are used extensively, especially in the analysis of intrastate toll demand. The tables in Chapters 3 and 4 containing estimates of price and income elasticities of demand provide a great deal of evidence that telephone demand is indeed a dynamic phenomenon, because when dynamic models are specified, the estimated long-run elasticities are nearly always considerably larger than the estimated short-run elasticities.

Perhaps the biggest problem with the treatment of dynamics in the empirical literature is that distributed-lag models are frequently forced to do too much. As we saw in Chapter 2, dynamic adjustment can arise from two sources: the first reflects the dynamics inherent in the consumption of a service generated by a complementary durable good,[7] while the second reflects inertia [of the type described in equations (106) and (109) of Chapter 2] that may exist in the short run.[8] A distributed-lag model can capture both types of dynamic processes, but the two processes cannot be separately identified. Much of the time, this cannot be avoided, since estimates of the stock of complementary durable goods either do not exist or else are of too poor quality to be used. But in the telephone industry, this is not the case, for the data on the stock of telephones are in general quite good. The access/use distinction can accordingly be modeled directly (thereby taking into account the dynamics associated with the consumption of a service generated by a durable good), and the distributed-lag model can focus exclusively on capturing the short-run inertia.

On the other hand, one can question (my discussion in Section 9 of Chapter 2 notwithstanding) whether the Koyck model, which is used in the vast majority of cases, may be too restrictive. The Koyck model postulates geometric decay in the distributed lag, and also

---

[7] In the present context, the complementary durable good is the telephone system, while the service generated by the durable good is the use of the telephone system.

[8] To dispel possible confusion, let me be more specific. The dynamics inherent in the access/use division represent the traditional distinction between the short run and the long run in the presence of a durable good. Suppose that the telephone system is in steady-state equilibrium (ignore the complications introduced by the access externality), and let this equilibrium be disturbed by an increase (say) in income. In the short run, there will be an adjustment in the number of calls that are made using the existing stock of telephones (and possibly also in average duration). In the long run, the stock of telephones may also adjust. However, it may be that in the short run (i.e., when the stock of telephones is fixed) there is a delay in adjusting the number of calls that are made to the higher income. This is the short-run inertia referred to in the second source of dynamic adjustment.

constrains each independent variable to have the same lag structure. Much of the time, these restrictions are probably unrealistic, but are resorted to because multicollinearity precludes meaningful estimation of separate lag structures.[9] However, other more flexible, distributed-lag models exist, and these should be analyzed with the purpose of seeing whether price and income have different lag structures. The Almon polynomial-lag model would be a convenient model to use to this end. Transfer functions, which have been used in a couple of studies with promising results, also merit consideration, particularly when quarterly or monthly data are being analyzed.[10]

Let me now turn to matters of a purely econometric nature. The discussion in this survey has touched on econometric and statistical problems only in passing. This is not to say that faulty practices are absent, but my view throughout this survey has been that the econometric problems are of a lower order or priority than the problems of model specification. In general, the econometric analysis, qua econometrics, is essentially state-of-the-art for the types of investigations that are involved. There are a few places where autocorrelation in the residuals appears to have been a problem, but this has been dealt with using one of the standard corrections (Cochrane-Orcutt, Hildreth-Lu). On the other hand, there are several studies where autocorrelation could be a problem, but the authors do not report whether this is the case. In most of the state models, first-order autocorrelation is tested for using the Durbin h-test, which is the appropriate test to use, since these models have the lagged value of the dependent variable as a predictor. In the state models for one Bell System operating telephone company, the correlogram of the residuals is calculated, which allows for higher order autocorrela-

---

[9] Another frequently heard criticism of the Koyck model is that its estimating equation contains the lagged dependent variable as a predictor, and is accordingly susceptible to the ills of bias and inconsistency. This is not the place to undertake a general critique of models of this type. Suffice it to say that I have used models involving lagged values of the dependent variables for many years, and will probably do so for many years to come. This is not to say that one should not exercise a great deal of caution with the Koyck model, especially in checking for the presence of autocorrelation in the error term. As is well known, positive autocorrelation can lead to a serious upward bias in the coefficient for the lagged dependent variable, which in turn will lead to implausibly long periods of adjustment and possibly to seriously biased estimates of long-run elasticities. The Durbin-Watson test, is not the appropriate test for autocorrelation in this situation, but rather the Durbin h-test. Also, one needs to be especially careful in estimating a Koyck model from a pooled cross-section/time-series data set.

[10] For a discussion of transfer functions, see Box and Jenkins (1976). Cf. also Fask and Robinson (1977).

tion to be checked for. This is particularly important in models that have been estimated with quarterly or monthly data.

One thing that has not been adequately exploited is the time-series/cross-section nature of much of the telephone data base.[11] The data collected in the CMDS would allow for this, and so, too, would the data collected at the state level by the operating telephone companies. The major benefit from pooling is the increased variation in the independent variables. However, care must be taken that the structures being pooled are homogeneous—that is, that regression coefficients are constant across observational units, and similarly for the structure of the error term. Traditionally, the covariance model has been the model that has been employed in pooled time-series/cross-section analysis, but recent years have seen an increasing use of the variance-components model. Random coefficients models may also have a role to play, especially in situations where price or income elasticities are found to vary regionally.[12]

Since many of the econometric models of telephone demand are used in rate filings, the basic canons of econometrics need to be given a great deal of attention, for otherwise the results of the models, or even the models themselves, may be challenged on the grounds that proper econometric and statistical procedures have not been followed. Much of the time, econometricians take many things for granted, especially when communicating with other econometricians; to do otherwise would be tedious, repetitive, and time-consuming. The validity of the t- and F-tests, for example, requires that the

[11] There are only six studies in the literature that estimate models using pooled time-series/cross-section data. Kearns (1978) and Reitman (1977) pool time-series data across states in estimating models of the demand for residence extensions (Kearns, Reitman) and the demand for total vertical services (Reitman). Stuntebeck (1976) and Wert (1976) both use pooled data in analyzing the daytime/nighttime composition of toll traffic. Finally, Deschamps (1974) uses a pooled time-series/cross-section data set in his analysis of toll demand in Belgium, and Rea and Lage (1978) do the same in their analyses of international telephone, telex, and telegraph demand.

[12] The standard econometric reference on the covariance model is Kuh (1963). See also Houthakker (1965). The variance-components model was developed by Balestra and Nerlove (1966). See also Madalla (1971), Wallace and Hassain (1969), and Nerlove (1971). For a discussion of random coefficients models, see Swamy (1971). Of the studies listed in Footnote 11, Deschamps and Rea and Lage are the only ones to use a variance-components framework. Kearns, Stuntebeck, and Wert use a covariance model, but Reitman pools directly without either the covariance or variance-components adjustment. The consequences of pooling directly are especially severe in Reitman's study, for a model is used in which the lagged value of the dependent variable is included as a predictor. For models of this type, a covariance or variance-components framework is necessary; otherwise, the individual state effects will be reflected in the lagged value of the dependent variable, and the estimate of its coefficient may be severely biased.

error term be normally distributed, but rarely is normality tested for explicitly. Usually, the Classical Central Limit Theorem is relied upon to provide normality, but failing this, the econometrician knows that the t- and F-tests are robust even in the face of quite substantial departures from normality. However, the people that ultimately have to be convinced in rate filings are not other econometricians, but rather people who may have little understanding or appreciation of econometric procedures. As a consequence, laxity and possible errors in procedure can be made to seem much more important than they in fact are. Hence, in using a model in a rate filing, formal procedures need to be followed with especial care, including (to return to the example) an explicit test for normality in the error term.[13]

SUMMARY OF THE EMPIRICAL RESULTS AND SOME IMPLICATIONS THEREOF

At several points in Chapters 3 and 4 I indicated some of the general relationships among the estimated price and income elasticities that we might reasonably expect to find. Common sense suggests that access to the system should be more of a necessity than local use and that local use should be more of a necessity than toll use. This would mean that the income elasticity for access would be smaller than the income elasticity for local use, which, in turn, would be smaller than the income elasticity for toll use. In general, we should expect the same relationships to hold among the price elasticities, although the reasoning is somewhat more subtle. The effect on demand of a change in price, to recall, consists of two terms, an income effect and a substitution effect. If the substitution effect is held constant, the price elasticities will in general, increase *pari passu* with the income elasticities. However, common sense also suggests that the substitution effect will be very weak for access, and weaker for local use than for toll use. Thus, both the income effect and the substitution effect should make for progressively larger price elas-

---

[13] I have singled out normality of the error term because it is usually not considered a trouble point. The conventional list of statistical problems includes autocorrelation, heteroscedasticity, and multicollinearity. Autocorrelation is nearly always a problem with aggregate time-series data, and extreme care should be exercised in checking for autocorrelation in situations where t-ratios are around 2. Heteroscedasticity, on the other hand, is more likely to be a problem with cross-section data than with time-series data. In general, there is porbably more laxity in checking for heteroscedasticity than in checking for autocorrelation. Finally, multicollinearity is nearly always present in some degree, no matter what the source of the data, although it is typically strongest with time-series data.

ticities as we move from the demand for access to the demand for local use, to the demand for short-haul toll, and so on.

The empirical results tabulated in Sections 2-4 of Chapter 3 and 1-2 of Chapter 4 support these views. In general, the estimated elasticities for access are smaller than the elasticities for local use, which are smaller than the elasticities for toll demand. This is true for both the income and the price elasticities. Moreover, the empirical results also indicate that the elasticities for toll demand vary with distance, being smaller for short-haul calls than for long-haul calls. The elasticities in the intrastate models are, in general smaller than in the interstate models, and the short-haul/long-haul pattern is also exhibited in the elasticities in the LDI model.

In Table 5-1, I have tabulated some point estimates of price and income elasticities for the demand for access, local calls, and long-distance calls and for the duration of long-distance calls. *These estimates are my own interpretation of the existing empirical record.* These estimates are highly subjective and are based on evidence from other countries as well as the United States. While the estimates tend to be near the mid-point of existing estimates, I have given some studies more weight than others. The estimates refer to steady-state, long-run elasticities, and are for residence and business demands combined, for the empirical record at this stage will not support an attempt to distinguish between residence and business customers. As a measure of the uncertainty to be associated with the estimates, I have appended a range to each estimate. *These ranges are also subjective, and primarily reflect my own views as to the intervals within which the true elasticities lie.*

The greatest uncertainty attaches to the estimates of the income elasticities of the demand for use, particularly for local calls and interstate toll calls. I think that one can conclude that the income elasticity for toll calls is greater than 1, but how much greater than 1 is still an open question. Considerable uncertainty also surrounds the price elasticity for interstate toll, especially with respect to the critical value of 1 (in absolute value). There is a strong feeling within the telephone industry that the price elasticity for interstate toll is less than 1, and indeed, is most likely in the neighborhood of 0.5. If we return to Table 3-5 of Chapter 3, we see that there are only three studies that seriously suggest that the long-run price elasticity for interstate toll may be as low as 0.5, the LDI model, Deschamps (1974), and Kwok, Lee, and Pearce (1975). However, Deschamps' study is for Belgium, which corresponds to very short-haul traffic in the context of the U.S. interstate toll market, and Kwok, Lee and Pearce find a much higher elasticity with the data for the Trans

Table 5-1. Point and Interval Estimates of Price and Income Elasticities of Demand for Selected Telephone Services

	Elasticity			
Type of Demand	*Service Connection Charge*	*Monthly Service Charge*	*Toll Price*	*Income*
Access	−0.03 (±0.01)	−0.10 (±0.09)	—	0.50 (±0.10)
Local calls	—	−0.20 (±0.05)	—	1.00 (±0.40)
Toll Calls (conversation-minutes)				
Intrastate	—	—	−0.65 (±0.15)	1.25 (±0.25)
Interstate	—	—	−0.75 (±0.20)	1.50 (±0.40)
International calls	—	—	−0.90 (±0.30)	1.70 (±0.40)
Duration of toll calls	—	—	−0.15 (±0.05)	0.25 (±0.10)

Source: These estimates refer to long-run, steady-state elasticities. The estimates for toll calls refer to conversation-minutes, rather than to just the number of messages. The estimates reflect my own interpretation of the empirical record (for both foreign countries and the United States) and are thus highly subjective. The numbers in parentheses are subjective standard errors.

Canada Telephone System than with the data of Bell Canada for just Ontario and Quebec. Thus, we are essentially left with the LDI model. As we saw in Section 4 of Chapter 3, though, the LDI elasticities have been arrived as through simulation techniques, rather than estimated econometrically, and it is possible that some of the effects of falling interstate toll rates during the 1960s were incorrectly attributed to general economic growth. At any rate, my own view at this juncture is that the long-run price elasticity for long-haul toll calls is as close to 1 as to 0.5.

The entries in Table 5-1 provides only a partial listing of the categories of telephone demand. I have not included any elasticities for WATS and private line, vertical services, and coin stations, and, as already mentioned, I have not attempted a residence/business breakdown. The empirical evidence in all of these areas is too weak for the tabulation of "best-guess" point estimates of elasticities. For WATS and private line, what evidence there is suggests that own-price and income elasticities are more or less the same as for MTS, and there is solid evidence in the Feldman study (1976) that WATS, private line, and MTS are strong substitutes. Yet, interestingly, there is a suggestion in the study of Subissati (1973) that WATS and MTS may be complements. What Subissati finds is that current expenditures for MTS are positively related to the first difference in expenditures for WATS. Subissati suggests that this may reflect a stimulus to total toll calling induced by the presence of an additional way to make toll calls. If this is in fact the case, then WATS, private line, and MTS may actually be complements in the long run, but substitutes in the short run. In any event, the WATS/private line/MTS tradeoff is an important area for future research.

The demand for calls from coin stations is also underresearched, and the same is true for the demand for vertical services. The Feldman study stands alone in econometric analysis of coin-station calls, but the focus is restricted to MTS calls and prices, so that no information is offered concerning the demand for local calls. In short, there does not exist any econometric evidence at the present time concerning the elasticity of demand for coin-station calls with respect to the price of a local call.[14] With regard to the demand for vertical services, most of the analysis has focused on the demand for residence extensions.[15] Kearns (1978), in particular, obtains a

[14] In reaching this conclusion, I have not had the benefit of reading a recent study of the demand for local calls from coin stations in Ontario and Quebec by Brubacher and Hiscock (1978).

[15] See Table 4-9 of Chapter 4.

price elasticity for residence extensions of −0.33 and an income elasticity of 0.25. Reitman (1977) obtains a price elasticity for total vertical services of −0.99, but I am skeptical of this estimate because (as discussed in Section 1) it was obtained from a pooled time-series/cross-section distributed-lag model that was estimated without allowing for individual state effects.

Let me now turn to some implications of the findings concerning the price and income elasticities of demand. To begin with, it should be clear that price elasticities exist—that is, contrary to the views of many, they are not zero. On the other hand, it does appear that except possibly for very long-haul toll calls, telephone demand is inelastic. The demand for access, in particular, appears to be very price inelastic, although again, let me repeat: this elasticity is different from zero. And there is some evidence that the "transient" component of inward and outward movement is quite sensitive to the level of the service-connection charge.

The fact that the price elasticity for local use is small means that the telephone companies can look to the local market as a place to recoup the revenues that are almost certain to be lost in the long-haul toll and vertical-services market as a result of competition. However, the key question is whether the price of access to residential customers is to be continued to be subsidized. In the past, the subsidization of residential access has been justified in terms of fostering universal service. However, the real issue would seem to turn on the access externality. If this externality exists, then optimal social pricing requires that the price of access continue to be subsidized,[16] if it does not exist, then the price of access should be set equal to marginal cost.[17] Once again, we are reminded of the importance of establishing whether the access externality exists and estimating its quantitative magnitude.[18]

Let us suppose, for now, that the access externality is unimportant and that the telephone companies set out to increase the price of

[16] By optimal social pricing in this contest, I mean a pricing policy that has the objective of maximizing the sum of consumers' and producers' surplus. See Zajac (1979).

[17] Because of the rate-of-return (or other regulatory) constraints, optimal social pricing might require that the price for access deviate from marginal cost in a way that depends upon the price elasticity of the demand for access. This is usually referred to as Ramsey-pricing after F.P. Ramsey (1927). See Baumol and Bradford (1970) and Zajac (1979).

[18] If the access externality is quantitatively significant and if the telephone companies cannot subsidize access from the local-use market because of competition, how the subsidy is to be financed becomes a very important social question.

access to the level of marginal cost. What might be expected to happen? My own view is that there would probably be a great deal of resistance by residential customers. I base this observation on the empirical evidence obtained by Perl (1978) showing the dependence of the price elasticity for access on the level of income and the level of the access price. The empirical results in question are presented in Table 3-2 in Chapter 3. The point in question has already been discussed (in Section 2 of Chapter 3), but it will be useful to remind ourselves of what is involved. Refer once more to Table 3-2 of Chapter 3. The elasticity with respect to the access price is seen to decrease with the level of income and to increase with the level of the monthly service charge. The access price elasticity is accordingly largest for low-income households facing a high monthly service charge. On the other hand, the price elasticity is very small for high-income households facing a low monthly service charge. Thus, one can conclude that most of the access price elasticity arises from the income effect, rather than from the substitution effect. However, the substitution effect, to recall, implies no loss in consumer welfare, whereas the income effect does. Consequently, as the price of access is moved toward marginal cost, I would expect consumer resistance, particularly on the part of low-income households living in areas where the monthly service charge is already fairly high.

Thus far in this review, the focus has been mostly on price elasticities. Income elasticities have been treated largely in passing. The reason for this is that for the most part, income elasticities are not controversial—that they are greater than zero in virtually all cases and are substantial in many is accepted without comment—but this is not the case with price elasticities, especially during a period of frequent (and sometimes large) rate increases. Income elasticities are very important to the telephone industry because they indicate how, holding prices, technology, and other nonincome determinants of demand constant, the industry will develop over time, which markets will require the most additional investment, and where additional revenues will accrue. In the markets where the aggregate income elasticity is greater than 1, the growth in revenues will be slower than the growth of the general economy. Of course, nonincome factors do not remain constant, so that revenue growth in individual markets can be quite different than that implied by income elasticities alone. In general, however, one should expect revenues to grow most rapidly in the markets with the highest income elasticities of demand.

As was noted earlier in this section, the estimates of income elasticities are in general quite large, although there is a lot of variation in the estimates, especially in the ones for local use and interstate toll.[19] In Table 5-1, I have suggested point estimates of at least 1 in all of the major telephone markets except for access. As noted, however, a great deal of uncertainty surrounds the estimate for local use, so that this elasticity can very well be less than 1. The income elasticity for toll calls, though, is almost certainly greater than 1, particularly in the long-haul interstate market. And there seems little question but what the elasticity for overseas calls is substantially in excess of 1. Indeed, for a number of years overseas revenues have been the fastest growing component of total Bell System profits, and, if present trends continue, there could be the day when they are the most important contributor.

PROBLEM AREAS

In the last two sections of this chapter, I shall discuss some of the more pressing problem areas in telephone demand and suggest a research program for dealing with them. Business demand should be near the top of the list. In Chapters 3 and 4, we found only a few empirical studies of business demand, and in Chapter 2, we discussed the absence from the literature of a coherent theoretical model of business telephone demand. In Section 8 of Chapter 2, we discussed some of the general features of business telephone demand, but there was no attempt to represent these in a formal theoretical structure. The absence of a business focus is particularly apparent in the extremely limited treatment in the literature of WATS and private-line demands and of the tradeoffs among WATS, private line, and MTS.

After business demand, much remains to be done in developing and extending the theoretical framework for residential demand. The empirical models for access and use need to be brought into closer agreement with the theoretical structure expressed in equations (53) and (64) in Chapter 2. However, these expressions represent complicated implicit functions, and it may not be possible to find a utility function that yields explicit closed-form expressions for the demand functions. But we will not know whether this is the case until the problem is thoroughly explored. Option demand also needs a more rigorous theoretical treatment, and empirical procedures need to be devised for quantifying its importance. Empirical procedures also need to be devised for quantifying the impacts of the access and call externalities, particularly the access externality.

---

[19] See Tables 3-4 and 3-5 in Chapter 3.

Indeed, in my opinion, the isolation of the access externality is the most immediately pressing empirical issue in telephone demand. For, as discussed earlier, a socially optimal policy for pricing access to the telephone network cannot be developed until the quantitative significance of the access esternality is resolved.

Still on the access externality, a question that needs to be explored is whether the access externality is invariant with respect to system size. The view that has been implicit throughout this monograph is that the access externality is positive, but this may not always be the case. What is more likely, it seems to me, is that the importance of the externality diminishes as the telephone system reaches a certain size, and it may be that the externality is actually negative for very large systems. The fact that large systems have larger proportions of subscribers with unlisted and nonpublished numbers than small systems provides some empirical support for this view. However, what really matters for the access externality is not system size per se, but the community of interest, for once a subscriber's community of interest is fully contained in the telephone system, the addition of more subscribers is largely irrelevant. Thus, further research on the access externality should include sharpening the concept of the community of interest and devising improved ways of measuring it.

Virtually all efforts in applied demand analysis face index number problems, and telephone demand is no exception. For several years, a group at AT&T has been constructing telephone price indexes for use throughout the Bell System.[20] The indexes have been devised for a variety of purposes, many of them unrelated to demand analysis, but have also been used extensively to that end. Price indices now exist for nearly every category of message toll service and inward and outward WATS, and work is currently underway to construct indexes for the major components of access and local service. The indexes are of a Laspeyre's, fixed-weight type and are reweighted every four years, except for the index for interstate MTS, which has been reweighted every two years since 1972. Data for the weights, which were originally obtained from the MASP, are now obtained from the CMDS.[21]

AT&T is to be applauded for its support of this price index work. The indexes have been constructed with care from a rich data base,

---

[20] See AT&T (1976b) for descriptions of the Price Index Project.

[21] For a description of the CMDS data, see the Proceedings in FCC Docket 78-302, in particular, "Notice of Proposed Rule Making," September 28, 1978, and AT&T's response, December 5, 1978, and meeting notes regarding FCC Docket 78-302, May 16, 1979. For a description of the MASP data, see AT&T (1978).

and have found extensive use. However, the indexes in their current form have some shortcomings, and one of my recommendations is that the whole focus of telephone price indexes be the subject of a separate review. I have three specific areas of concern:

1. *Multi-part tariffs.* The AT&T indexes do not deal with multi-part tariffs in a satisfactory way. The price of toll calls, as has been mentioned at several points, should be represented by two indexes, one for the initial-period charge and the other for the overtime-period charge. At present, there is a single index that measures the price of a composite MTS call. The price index of local service also needs to be disaggregated. Service-connection charges, basic service, and vertical services should all have their own index (or indexes, when multi-part tariffs are involved), and where local service is measured service, there should be separate indexes to reflect the various dimensions of the measured service. In addition, there should be separate indexes for residence and business customers. However, in fairness to the AT&T Price Index Project, I should say that price indexes for demand modeling should be developed in response to the needs of demand models, rather than demand models being accommodated to the availability of price indexes. (At least this should be the case in the long run.) And this is especially true in the telephone industry where the data base is in general good and appropriate indexes can, for the most part, be constructed.
2. *Aggregation problems.* Most of the empirical work on telephone demand has been at high levels of aggregation. Numerous classes of service are aggregated in defining local service, and mileage bands are aggregated in defining toll demand. At the present time, aggregation proceeds in the AT&T toll-price indexes through the use of base-period MTS call profiles as weights. Duration is taken into account by weighting each cell in the profile by the average duration of calls in the cell during the base period. A number of questions can be raised about this implicit aggregation procedure, but an area of particular concern involves the dependence of price on mileage in price indexes that aggregate over mileage bands.[22]
3. *Reweighting.* The AT&T indices are currently reweighted at four-year intervals, except for the interstate MTS index, which is reweighted every two years. However, when the structure of the rate schedule, as well as the level of the schedule, is changed

[22] See Gale (1974, pp. 8-9) for one attempt to deal with this question.

frequently, as has been the case in recent years, frequent reweighting is desirable. The reason for this is that the "market basket" for telephone use will be changing, and these changes should be reflected in the weights. Indeed, Paasche indexes, or even chain indexes, warrant consideration.[23]

The mention of aggregation problems brings up a final point, namely, the need for greater disaggregation. We have already emphasized this with regard to decomposing toll traffic into the number of calls and average duration, distinguishing between residence and business customers, and so on. Business customers, as was discussed in Section 8 of Chapter 2, ought to be grouped according to the type of economic activity that the business is engaged in. Telecommunications needs clearly vary substantially across types of economic activity, and disaggregation is required in order to take this into account, although, unfortunately, it appears that existing data bases can support this on only a limited basis. Point-to-point modeling of business demand is also warranted, particularly to get at the tradeoffs among WATS, private line, and MTS. But here, too, it appears that the data necessary to do this may not exist.[24]

Greater disaggregation is also desirable for residential customers. Several microdata sets, consisting of observations on individual households, already exist, and additional ones are almost certain to be generated in conjunction with the movement to measured local service. Also, as was discussed in Section 1 above, the data that exist at the state and local exchange levels have not yet been fully exploited. In particular, I think that much can be learned about the access externality using cross-section data for local exchanges. And, finally, the study of Perl (1978) illustrates well the benefits from using data from the Census of Population.

SOME FINAL COMMENTS AND AN AGENDA FOR FUTURE RESEARCH

In this section, I shall bring this long review to a close by presenting some concluding observations regarding the present state of demand

[23] As was mentioned in Chapter 4, chain indexes have been employed in eleven state toll demand models. In discussions with Bell Canada analysts, they have indicated that price indexes are also an area of continuing investigation at Bell Canada.

[24] Limited data exist for a number of the largest business customers of the Bell System, but there is a problem in using these data in building models of point-to-point demand. For large corporations with many establishments, there may be little relationship between the use that is billed to a location and the actual use at that location.

analysis in the telephone industry and a list of recommendations as to where to go from here. On the whole, the quality of the analysis in the empirical literature is good. The models that have been analyzed are essntially state-of-the-art for applied demand analysis, and this is also true for the econometric techniques that have been used in estimation. The empirical literature contains a number of good studies and several really excellent ones. Included among the latter are Deschamps (1974), Gale (1974), Feldman (1976), Larsen and McCleary (1970), Pavarini (1975, 1976b, 1979), Perl (1978), Pousette (1976), Stuntebeck (1976), and Waverman (1974). The theoretical literature also contains some first-rate contributions, with the list being headed by Artle and Averous (1973), Littlechild (1975), Rohlfs (1974), and Squire (1973).[25]

Still, the quality of the empirical literature falls short of where it could be. The biggest problem is that the empirical and theoretical analyses of telephone demand have been like two ships passing in the night. The best empirical work has ignored the best theoretical work, and vice versa. Clearly, the theoreticians and the applied analysts need to become aware of each other's activities—or, even better, join forces. Telephone demand modelers would also benefit from greater contact with the experience of demand analysts in other areas, particularly energy demand. Pooled time-series/cross-section models have been used extensively in analyzing energy demand, and the experience that has accumulated there is clearly relevant to the increased use of similar models for telephone demand. Energy demand analysts have also had extensive empirical experience in dealing with the problems caused by multi-part tariffs.[26]

When we look at the empirical literature as a whole, toll demand in comparison with access and local service demand, toll demand has received a disproportionate share of attention. The demand for vertical services is also under-researched, and WATS and private line (particularly private line) have been practically ignored. International demand has received some attention, but not as much as its rapid growth warrants. Finally, residential demand has received considerably more attention than business demand. That toll demand has been the center of attention is readily understandable, for toll demand is relatively easy to model, and rate activity in recent years

[25] I view the studies cited in this paragraph as constituting a "bare-bones" reading list for anyone wishing to become familiar with the literature on telephone demand. Two older studies are also highly recommended, namely, Krapelian (1958) and Luenbach (1958).

[26] See, for example, Taylor, Blattenberger, and Verleger (1977) and Acton, Mitchell, and Mowill (1976).

has tended to concentrate on the toll markets. However, the research focus needs to change. While there is need for continued research on toll demand, the area most in need of attention is the demand for local use, followed by the demand for access, the demand for WATS and private line, the demand for vertical services, business demand, and the demand for coin stations. And, as noted earlier, with regard to toll demand, the area most in need of additional attention is point-to-point demand. For now that there is competition in the intercity toll and private line markets, route-by-route pricing of calls becomes likely, and knowledge of point-to-point price elasticities is of obvious relevance.

In Chapters 3 and 4, efforts were made to tabulate the existing estimates of price and income elasticities of demand. Although we cannot take these collections of estimates as definitive, a solid conclusion emerges: *price elasticities of demand are different from zero.* Yet to state that price elasticities are different from zero is not to say what their values actually are. A great deal of uncertainty surrounds virtually all of the estimates, and one of the major tasks of future research is to reduce the zones of uncertainty. There is some evidence that the price elasticity for long-haul toll calls may be as large as -1, and since -1 is a decidedly critical value, it is particularly important that the uncertainty associated with the price elasticity for toll demand be reduced. The efforts to do this, however, should distinguish more than has been the case in the past between the number of calls and the duration of calls, and price indexes should be used that capture the essential characteristics of multi-part tariffs.

There is also considerable uncertainty surrounding the existing estimates of income elasticities, especially the income elasticities for local calls and long-haul toll calls. However, despite the uncertainty of the estimates, the income elasticities appear to be at least 1 in all of the major markets except for access, and well in excess of 1 in the long-haul toll and international markets.

Finally, let me make a few comments regarding data. The telephone industry would appear to have one of the best data bases in existence for undertaking empirical demand analysis. Merely to describe it would require a major study in its own right. Yet, given the many facets of telephone demand and the complexities of the tariff structure, the existing data base needs to be expanded and improved. For, while very good data exist for some services, there are other services for which this is not the case. Coin stations, WATS and private line, and vertical services are cases in point. I have described some of the improvements that are needed in connection with price indexes, and I also have touched briefly on the data needed to estimate a model for business demand. In some cases, the data now exist for overcoming the weaknesses in the existing literature—although the organization of the data may be both time-consuming and expensive. However, this does not appear to be the case with regard to examining some of the more pressing issues in local-service demand, estimating models of the demand for terminal equipment. devising a comprehensive model of point-to-point toll demand, or estimating a disaggregated model of business demand.

My final recommendation concerns something that has been outside of the purview of this survey, namely, costs. At a number of points, I have noted that there are forces in motion that will probably cause the telephone industry to abandon its traditional pricing philosophy based on value of service, residual pricing of basic service, and so on in favor of a policy that is more closely tied to the marginal cost of providing services. If this occurs, the structure of demand will be only one side of the pricing coin; the structure of costs will be the other side. Accordingly, my last recommendation is that a study, along the lines of the present one, also be undertaken for costs.

Chapter 6

Postscript

The cutoff date for most of the studies reviewed in this monograph was April 1978. However, telecommunications demand is now very topical, and a lot has happened since the first draft of this monograph was completed. Most of the intrastate toll demand models tabulated in Chapter 4 are already out of date, new studies have appeared, and there is a great deal of ongoing research—much of it, I am pleased to say, in line with the suggestions made in the last chapter. The purpose of this postscript is to describe briefly a couple of studies that I had previously overlooked and to discuss work currently in progress at AT&T Long Lines that has the purpose of developing a model of U.S. interstate toll demand to replace the LDI model.

The two studies that had escaped my earlier attention are by Irish (1974) and Mahan (1979). The focus in both studies is North Carolina. The unit of observation in Irish's study, which was undertaken when he was on the staff of the North Carolina Utility Commission, is the local exchange, while the unit of observation in Mahan's study, which was his Ph.D. dissertation in the Department of Economics at North Carolina State University, is an individual subscriber. The data used by Irish are for all of North Carolina for the years 1968, 1970, and 1972, while the data used by Mahan refer to a sample of subscribers in eastern North Carolina served by Carolina Telephone and Telegraph Co. in 1975. Since Mahan's data are more current, I shall direct most of my attention to his study. However, Irish's study is also recommended to readers because of the effort to take into account the number of telephones that can be reached in the free-calling area in explaining the cross-sectional variation in the number of telephone subscribers across local exchanges. In doing this, Irish was attempting to deal directly and explicitly for the access externality. His findings are that the local calling scope is an important determinant of access demand, especially in the exchanges outside of the areas served by Southern Bell, which tend to be smaller, rural exchanges.[1]

Mahan's study involves an analysis of the factors determining a household's decision to subscribe to the telephone network, its choice of grade of service, its decision to rent an extension telephone, and its expenditures for toll calls. The explanatory factors include economic variables such as income and price, sociodemographic variables such as household size and composition by age, degree of urbanization, race, occupation, and education, and special factors such as geographical dispersion of family, size of housing unit, and calling scope. The geographical dispersion of the family and calling scope (which is measured by the number of main stations that can be reached within the local calling area) represent measures of the household's community of interest.

A traditional double-logarithmic demand function is specified for expenditures for long-distance calls, while binary probit models are used for the decision to purchase access, the choice of grade of service, and the decision to rent an extension. Cross-sectional data, consisting of individual observations on a sample of households in eastern North Carolina, are used in estimation. From mid-March to early June 1975, the Department of Economics and Business of North Carolina State University completed interviews, as a part of a survey of labor supply in eastern North Carolina, with a sample of 2,113 households in eight eastern North Carolina counties. The socioeconomic data from these interview were combined with the telephone records for the same households from Carolina Telephone and Telegraph Company. Of the 1,601 households (out of the 2,113 households) living in single economic family housing units and having telephone service, 1,138 gave written permission for the telephone company to release records on the telephone equipment installed in their homes, as well as detailed information on long-distance calls placed from their telephones.

---

[1] Irish bases his analysis of the accesss externality on the theoretical work of Artle and Averous (1973) and Squire (1973). [The paper of Rohlfs (1974) had probably not yet appeared.] Irish's study is thus counter to my statement in Chapter 5 that the best theoretical work on telephone demand has not found its way into the econometric literature. Irish also has some very sensible things to say about the determinants of business telephone demand, and in this regard, too, his study is pioneering.

This is the sample that was used in estimating the demand for long-distance calls, the decision to rent an extension telephone, and the choice of class of service. A smaller sample of 863 households was used in estimating the demand for access. (Of the 863 households, 691 [or 79 percent] had a telephone at the time of the survey.)

The Mahan study is an important addition to the short list of studies that use data at the household level, and, as far as I am aware, is the first to approach the choice of class of service in a sequential qualitative choice framework. Mahan's procedure is, first, to explain the decision to have a telephone (irrespective of the grade of service), then, of the households with telephones, to explain the proportion that selects one-part service, and then, of the households that choose two- or four-party service, to explain the proportion that selects two-party service. A binary probit model is used at each step in the sequence.

Mahan's results can be summarized briefly as follows. The level of income has an important effect on the probability of a household subscribing to telephone service, on the prbability of renting an extension telephone, and on the expenditures for long-distance calls. The estimated income elasticities for telephone service, extension telephones, and toll calls are 0.15, 0.21, and 0.17, respectively. However, the level of income is not important in explaining the choice of class of service. With the cross-sectional data analyzed, the only variation in price is in the monthly service charge for main stations, but [unlike for Perl (1978)] this variable is insignificant in explaining a household's decision to subscribe to the network. On the other hand, the rates for the different grades of service are found to have a significant impact on the choice of grade of service.

Mahan also finds that many noneconomic factors have a significant impact. Rural households are less likely to rent an extension telephone, and they spend less on long-distance calls than similarly situated urban households. Black households are less likely to have a telephone, but those who do are more likely than whites to select a higher grade of service, rent an extension telephone, and spend more on toll calls.[2] The educational level of the head of the household has a significant positive effect in each of the categories of demand considered. Finally, long-distance calling increases with the number of household members eighteen years of age or older

---

[2] This last finding, it will be noted, is somewhat at odds with what was found in the Brandon study (forthcoming) for black households living in Chicago. In the Chicago study, black households were found to make more local calls than white households, but fewer (toll) calls to the suburbs.

and with increases in the geographical dispersion of the household's immediate relatives. On the other hand, toll calling diminishes with increases in the size of the exchange in which the household is located. This last result is of interest because it suggests that there may be an upper limit to the size of a household's community of interest.[3]

Let me now turn to an ongoing research effort at AT&T Long Lines which has the purpose of developing an up-to-date alternative to the LDI model for use in evaluating rate changes. The project, now known as the Rate Evaluation System (RES), was initiated in mid-June 1974 in accordance with an agreement between the FCC and AT&T Long Lines, and was conceived originally as a joint effort between the FCC staff and Long Lines Business Research. The RES model will be used in place of the LDI model in evaluating the revenue effects of changes in the interstate rate schedule. Another modeling effort at Long Lines is the Forecasting Interstate Revenues and Messages (FIRM) model, which is now being used to forecast the growth in domestic (forty-eight contiguous states plus Washington, D.C.) interstate messages for budget, financial planning purposes, and as input into the RES model.

The RES model is intended to model MTS demand at as disaggregated a level as possible. Equations are estimated (for both messages/telephone and average revenue/message) for fifty market segments within each of the forty-eight contiguous states plus Washington, D.C. These fifty market segments represent subdivisions by business and residence customers, by type of call, by rate period (for DDD only), and by five mileage bands. The model is estimated with a pooled time-series/cross-sectional data set which consists of quarterly observations on states. The time period currently covered by the sample is from the third quarter of 1971 through the second quarter of 1977. The data for the dependent variables are taken from the CMDS, and refer to Bell-billed messages from a state to the other forty-seven contiguous states and Washington, D.C.

A static double-logarithmic function is specified for the demand function for MTS messages per telephone.[4] The independent vari-

[3] Cf. the discussion of the increased importance of unlisted and nonpublished numbers in large local-calling areas in Section 3 of Chapter 5; cf. also footnote 20 in Chapter 2.

[4] Besides the double-logarithmic specification, linear and semilog models are also estimated. The "best" specification was then selected on the basis of a modified Box-Cox procedure which evaluates the log-likelihood function.

ables are an index of real interstate MTS own price, an index of real interstate competitive prices, and a measure of the level of economic activity (real labor income per employee in the business equations, and real disposable income per capita in the residence equations). Each state has its own separate intercept, and the models have been estimated using a generalized-least-squares technique that corrects for first-order auto-correlation within a state and for heteroscedasticity across states.

In certain markets, a constrained estimation procedure was used in order to overcome multicollinearity between the own and competitive prices. The particular procedure used is referred to as "sign-only" constrained estimation, and imposes the restrictions that the own-price elasticity is negative, the cross-price elasticity is positive, and the sum of the two elasticities is negative. This method of estimation was imposed only in those markets where the unrestricted models gave own-or cross-price elasticities of the wrong sign or, if the signs were right, the sum of the elasticities was positive. In general, the results obtained with this procedure lead to an improvement in terms of theoretically correct signs, but without any noticeable deterioration in the explanatory power of the equations.

The estimated net own-price elasticities of demand[5] in general increase (in absolute value) with the length of the haul,[6] which confirms earlier findings, are larger during the day than during the evening or night, and are larger for the residence markets than for business. The aggregate net own-price elasticities for business and residence using three different techniques of estimation are given in Table 6–1.[7] The first technique refers to the unconstrained log-linear model, while the second technique refers to the log-linear model estimated by "pooling over lamda." This technique imposes the sign-only restrictions mentioned earlier, but the restrictions are applied uniformly across markets by equalizing the ratio of the cross elasticity to the own elasticity across markets. Finally, the third technique refers to the "preferred" functional form (linear, log-linear, or semilog) as selected by the modified Box-Cox procedure.

Depending upon the estimation technique and whether for business or residence, the estimated aggregate net price elasticities are seen to be in the range of −0.40 to −0.65. These estimates, it will

[5] The net elasticity in this context refers to the sum of the own and cross elasticities.

[6] This relationship is not quite uniform across rate periods.

[7] The aggregate elasticity is calculated as a weighted average of the elasticities in the individual markets using messages as weights.

Table 6-1. Estimated Aggregate Net Price Elasticities, Interstate Messages/ Telephone (AT&T Long Lines Res Model)

Estimation Technique	*Business*	*Residence*
Unconstrained (log-linear)	−0.63	−0.64
Pooling over Lamda (log-linear)	−0.42	−0.53
Box-Cox preferred	−0.45	−0.46

Source: Business Research Organization, AT&T Long Lines.

be noted, while larger than the aggregate price elasticity implicit in the LDI model, are nevertheless smaller than most of the estimates tabulated in Table 3-5 in Chapter 3. This raises an interesting point, for the present estimates are calculated from a highly disaggregated model, whereas (with exception of the LDI model) the estimates tabulated in Table 3-5 of Chapter 3 are based on aggregate data. Consequently, an obvious question is whether concentration on aggregate data results in estimates of the aggregate price elasticity for long-distance calls that are biased upward. This question needs to be investigated.[8]

The RES model also includes equations for the length of conversation. However, the analysis at this stage is very preliminary. Equations have been estimated for the average number of overtime periods (ANOP), the average length of conversation (ANOC) and the average revenue per message (ARPM). The initial ANOP models were tested with both an index of the initial-period price and an index of the overtime-period price. As was suggested in Section 5 of Chapter 2, the overtime price should be most relevant in explaining duration, but in this case the initial-period price is included in order to take into account the shift in the initial period from three minutes to one minute. In the ALOC equations, the initial- and overtime-period prices are combined in a single index.

The results for the ANOP specifications have so far not been very encouraging, particularly concerning the sign for the initial-period price in the medium and high milage bands. The results for the ALOC specifications are more encouraging, but the best to date have been obtained for the ARPM specifications. For business

[8] The demand functions estimated in the RES model are static. Dynamic specifications, of the logarithmic Koyck variety, have also been estimated, but the lagged dependent variable is the only consistently significant predictor, and prices and income are usually insignificant. It is planned to investigate alternative dynamic specifications, but for now it is felt that the static specifications capture the most important features of interstate toll demand.

customers, the aggregate price elasticity for the average length of conversation is estimated to be −0.07, while for residential customers, the estimate is −0.17.

Let me now describe briefly the FIRM model, which includes models for forecasting both changes in messages and changes in length of conversation. Only the message model will be discussed. The dependent variable in the analysis is year-over-year percentage changes in interstate MTS messages. October MASP data are used from 1954 to 1966, while March and October MASP data are used from 1967 through March 1971. Finally, monthly data from the CMDS are used from July 1971 to June 1977. The message model contains equations for eleven markets, consisting of five mileage bands for both business and residence and one equation for all mileage bands combined for messages from coin stations. The models are estimated with both the dependent variable and the independent variables measured as year-over-year percentage changes. The eleven equations are estimated jointly using the Zellner estimator for seemingly unrelated regressions. A correction for first-order autocorrelation in the error term is applied to each equation. The estimates of the aggregate elasticities for price and income are given in Table 6-2.

The price elasticities, it will be noted, are considerably smaller than those reported for the RES model in Table 6-1. The income elasticities, on the other hand, are quite a bit larger than the ones obtained in the RES model. The rather small price elasticities in this model may be the result, at least in part, to their being estimated in a first difference model which contains an intercept. The intercepts in the equations, all of which are positive and highly significant, correspond to a linear trend in a levels equation, so that much of the effect of an almost continually decreasing real price of MTS calls over the period of the sample may be reflected in these intercepts.

Table 6-2. Estimated Aggregate Price and Income Elasticities (AT&T Long Line Firm Model)

Market	*Price Elasticity*	*Income Elasticity*
Business	−0.14	1.47
Residence	−0.19	0.63
Coin	−0.65	0.43
Total	−0.17	1.01

Source: Business Research Organization, AT&T Long Lines.

As noted at the outset of this postscript, telecommunications is now a very pervasive topic. With the rewriting of the Communications Act of 1934, the emergence of competition in the toll and terminal-equipment markets, and the consequent movement toward measured local service, conferences and workshops on the telecommunications industry are frequent, and almost all of these have at least some focus on telecommunications demand. In the preceding paragraphs, I have given a rather detailed description of work currently in progress at AT&T Long Lines. But the work at Long Lines is only the top of an iceberg. Research on telecommunications demand is ongoing at many other places: at AT&T, at Bell Canada and the Trans Canada Telephone System, at General Telephone and Electronics in Stamford (where the data being generated in the GT&E local measured service experiment in central Illinois are being analyzed), at the Rand Corporation, and even at the University of Arizona! There is no doubt in my mind that five years from now, nearly all of the results reported in this monograph will be obsolete. But such is progress.

Bibliography

Acton, J.P.; Mitchell, B.M.; and Mowill, R.S. (1976). "Residential Demand for Electricity in Los Angeles: An Econometric Study of Disaggregated Data," Rand Corporation (R-1899-NSF), Santa Monica, California, September.

Adelson, J.; Cannell, C.; Heyns, R.W.; and Lansing, J.B. (1956). "Motives, Attitudes, and Long-Distance Calls," Survey Research Center, Institute of Social Research, University of Michigan, April.

Aitchison, D., and Brown, J.A. (1957). *The Lognormal Distribution.* Cambridge, England: University Press.

Alleman, J.H. (1977). "The Pricing of Local Telephone Service," U.S. Department of Commerce, Office of Telecommunications, OT 77-14, April.

American Telephone and Telegraph Company (1975-76). "A Simulation of the Bell Revenue Loss and MCI Revenue Gain Based on Total Market Potential," unpublished paper, prepared for FCC Docket No. 20640.

____ (1976a). Affidavit of P. Muench, re: Execunet Service, FCC Docket 20614, August 18.

____ (1976b). Price Index Project, Comptrollers Department, December 10.

____ (1978). *1978 Annual FDC (Fully Distributed Costs) Report,* filed with the FCC.

Artle, R., and Averous, C. (1973). "The Telephone System as a Public Good: Static and Dynamic Aspects," *Bell Journal of Economics and Management Science* 4 (Spring): 89-100.

AT&T-Long Lines (1977). "Conclusive Evidence that Demand and Circuit Growth are Exponential," FCC Docket No. 18875, Section 7, pp. 1-58, August 31.

____ (1978a). "Demand Analysis and Forecast of the Mexico-U.S. Telecommunications Market," prepared by AT&T-Long Lines Overseas Economic Research and Telmex-Rates and International Planning, January (issued annually since 1976).

____ (1978b). "Demand Analysis and Forecast of the U.S.-Canada Telecommunications Market," prepared by AT&T–Long Lines Overseas Economic Research and TCTS-Network Development Forecasts, January (issued annually since 1976).

Auray, R.R. (1978). "Customer Response to Changes in Interstate MTS Rates." *Assessing the Impact of New Pricing Concepts in Public Utilities.* Harry Trebing, ed. East Lansing, Michigan: MSU Public Utilities Studies, Institute of Public Utilities, Michigan State University.

Balestra, P., and Nerlove, M. (1966). "Pooling Cross-Section and Time-Series Data in the Estimation of a Dynamic Model: The Demand for Natural Gas," *Econometrica* 34, (July): 585–612.

Baughman, M., and Joskow, P. (1975). "The Effects of Fuel Prices on Residential Appliance Choice in the U.S.," *Land Economics* 51 (February): 41–49.

Baumol, W.J., and Bradford, D.F. (1970). "Optimal Departures from Marginal Cost Pricing," *American Economic Review* 60 (June): 265–283.

Beauvais, E.C. (1977). "The Demand for Residential Telephone Service under Non-Metered Tariffs: Implications for Alternative Pricing Policies," paper presented at the Western Economic Association Meetings, Anaheim, California, June.

Bell Canada (1976). "Models of Message Toll Demand in Trans-Canada, Canada-U.S. and Canada-Overseas Settlements," Demand Analysis Group, Ottawa, Ontario, Canada, December.

Bell Telephone Company of Pennsylvania (1976). Testimony and Exhibits of Hendrik S. Houthakker, Dkt. No. 367, November 1976 and Cross-examination of June.

Bell Telephone Laboratories (1977). *Engineering and Operations in the Bell System*, Murray Hill, N.J.: Bell Telephone Laboratories.

Benedetti, M., et al. (1974). "Planning Concepts in Providing Telecommunications Services," *Gas/5 Manual*, Chapter XIII, International Telephone and Telegraph Consultative Committee, September.

Betteridge, W.W. (1973). "Description of Interstate Message Telecommunications Demand Model — Supplement," FCC Docket 19129, Phase II, November 1973.

Bhattacharya, M.N. (1974). "Forecasting the Demand for Telephones in Australia" *Applied Statistics* 23, No. 1: 1–10.

Black, S.K., and Tryon, P.V. (1976). "Increased Telephone Installation Rates: A Statistical Analysis of Colorado's 1972 Rate Change," U.S. Department of Commerce, OT Report 76-90, June 1976. Edited version in *Business Communications Review*, November/December.

Blattenberger, G.R. (1977). "Block Rate Pricing and the Residential Demand for Electricity," Ph.D. Dissertation, Department of Economics, University of Michigan.

Blomquist, N. (1977). "On the Relation between Change and Initial Value," *Journal of the American Statistical Association* 72, (December): 746–749.

Bø, Kristian; Gaustad, Olav; and Kosberg, Jan-Erik (1976). "Some Traffic Characteristics of Subscriber Categories and the Influence from Tariff Changes," paper presented at the Eighth International Tele-traffic Congress, Melbourne, Australia, November.

Box, C.E.P., and Jenkins, G.M. (1976). *Time Series Analysis Forecasting and Control.* San Francisco: Holden-Day, Inc., Revised Edition.

Brandon, B.B., ed. *The Effect of Demographics of Individual Households on Their Telephone Usage in Chicago.* Forthcoming.

Brooks, J. (1976). *Telephone: The First Hundred Years.* New York: Harper & Row.

Brubacher, S.R., and Hiscock, R.J. (1978). "An Application of Multivariate Box-Jenkins Techniques in Bell Canada," paper presented to the Joint ORSA/TIMS Conference, New York, May, Management Sciences Division, Bell Canada, Montreal.

Chen, G.K.C., and Winters, P.R. (1966). "Forecasting Peak Demand for an Electric Utility with a Hybrid Exponential Model," *Management Science* 12 (August): B 531–B 527.

C&P Telephone Company of Maryland (1977a). Rebuttal testimony of H. Kelejian, Docket No. 7025, March 29.

—— (1977b). Rebuttal testimony of A.P. Sprinkel, Docket No. 7025, March 29.

C&P Telephone Company of West Virginia (1977c). Testimony of Alan P. Sprinkel, Docket No. 8890, March 3.

Cohen, G. (1976). "Experimenting with the Effect of Tariff Changes on Traffic Patterns," paper presented at the Eighth International Teletraffic Congress, November.

Collins, S.M., and Infosino, W.J. (1978). "A Model to Estimate Cincinnati Local Residential Calling Rates from Demographic Variables," unpublished Bell Laboratories Memorandum, January.

Communications Satellite Corporation (1969). "Econometric Model of Demand for International Telecommunications," Washington, D.C.: Comsat Satellite Corporation, April.

Crane, D.B., and Crothy, J.R. (1967). "A Two-State Forecasting Model: Exponential Smoothing and Multiple Regression," *Management Science* 13 (April): B 501–B 507.

Craver, R.F. (1975). "International Calling at Bargain Rates," *Telephony* 189 (December 15): 34–35.

—— (1976a). "A Demand Analysis Study of Telecommunications Between Canada and the United States," unpublished paper, AT&T–Long Lines Over-Seas Economic Research, January and April.

—— (1976b). "A Look into a Crystal Ball at the Future of International Calling," *Telephony* 191 (August 2): 35–37.

—— (1976c). "An Estimate of the Price Elasticity of Demand for International Telecommunications," *Telecommunications Journal* 43, (November): 671–675.

—— (1976d). "UK Forecast Analysis," unpublished paper, AT&T–Long Lines Overseas Economic Research, September.

—— (1977). "The 'Utility' of Econometrics — How to Forecast Demand," *Telephone Engineer and Management,* (February 15): 74–78.

Davis, B.E.; Caccappolo, G.J.; and Chaudry, M.A. (1973). "An Econometric Planning Model for American Telephone and Telegraph Company," *Bell Journal of Economics and Management Science* 4 (Spring): 29–56.

Deschamps, P.J. (1974). "The Demand for Telephone Calls in Belgium, 1961-1969," paper presented at the Birmingham International Conference in Telecommunications Economics, Birmingham, England, May.

____ (1976). "Second-Best Pricing with Variable Product Quality: A Quantitative Decision Rule and an Existence Proof," Université Catholique de Louvain, Belgium.

____ (1977). "Pricing for Congestion in Telephone Networks: A Numerical Example," Center for Operations Research and Econometrics, Université Catholique de Louvain, Belgium.

Dobell, A.R.; Taylor, L.D.; Waverman L.; Liu, T.H.; and Copeland, M.D.G. (1970). *Communications in Canada: A Statistical Summary*, study prepared for the Department of Communications, Ottawa, by the Institute for Policy Analysis, University of Toronto, September.

____ (1972). "Telephone Communications in Canada: Demand, Production, and Investment Decisions," *Bell Journal of Economics and Management Science* 3 (Spring): 175-219.

Doherty, A.N. (1978). "Econometric Estimation of Local Telephone Price Elasticities," *Assessing the Impact of New Pricing Utilities.* Harry Trebing, ed. East Lansing, Michigan: MSU Public Utilities Studies, Institute of Public Utilities, Michigan State University.

____, and Oscar, G.M. (1977). "Will the Rates Produce the Revenues?" *Public Utilities Fortnightly* 99 (May 12): 15-23.

Domencich, T., and McFadden, D. (1975). *Urban Travel Demand: A Behavioral Analysis.* Amsterdam: North-Holland.

Dunn, D.M.; Williams, W.H.; Spivey, W.A. (1971). "Analysis and Prediction of Telephone Demand in Local Geographical Areas," *Bell Journal of Economics and Management Science* 2 (Autumn): 561-576.

Fask, A., and Robinson, P.B. (1977). "The Analysis of Telephone Demand by Dynamic Regression," unpublished paper, AT&T, September.

Feldman, J. (1976). "A Preliminary Cross Sectional Analysis of Services," unpublished paper, AT&T, February.

Fenton, C.G., and Schankerman, M.A. (1977). "The Demand for Basic Exchange Service: A Discrete Choice Model and Its Implications for Socially Preferred Subsidy Patterns," paper prepared for the Fifth Annual Telecommunications Policy Research Conference, Airlie, Virginia, March 30-April 2.

Finney, D.J. (1971). *Probit Analysis, Third Edition.* Cambridge, England: Cambridge University Press.

Gale W.A. (1971). "Duration of Interstate Calls, March 1969," unpublished Bell Laboratories Memorandum, December.

____ (1973). "Time of Day Distributions for Long Distance Calls," unpublished Bell Laboratories Memorandum, October.

____ (1974). "Elasticity of Duration for Intrastate Calls," unpublished Bell Laboratories Memorandum, October.

Garfinkel, L. (1977a). "Practical Problems in Economic Theory and Rate Structure," published in *Proceedings of 1977 Symposium on Problems of Regulated Industries*, February.

_____ (1977b). "Planning for a Transition in Pricing Basic Exchange Service," paper presented at Public Utilities Forecasting Conference, BOWNESS-ON-WINDERMERE, United Kingdom, March 24.

_____. "Network Access Pricing—Practical Problems in Implementation," *Issues in Public Utility Regulation.* H.M. Trebing, ed. East Lansing, Michigan: Institute of Public Utilities, Michigan State University, forthcoming.

Goldberger, A.S. (1964). *Econometric Theory.* New York: J. Wiley & Sons.

Granger, C.W.J., and Newbold, P. (1977). *Forecasting Economic Time Series.* New York: Academic Press.

Green, P.E.; Carmone, F.J.; and Wachpress, D.P. (1976). "On the Analysis of Qualitative Data in Marketing Research," working paper, American Telephone and Telegraph Company, May.

Harbus, F.I. (1975). "Comparison of Average Durations and Time of Day Distributions for WATS and Business MTS Calls," unpublished Bell Laboratories Memorandum, August.

Hartman, M.G. (1976). "Customer Line Usage Studies," paper presented at the Eighth International Teletraffic Congress, Melbourne, Australia, November.

Hausman, J.A., and Wise, D.A. (1978). "A Conditional Probit Model for Qualitative Choice: Discrete Decisions Recognizing Interdependence and Heterogenous Preference," *Econometrica* 46 (March): 403–426.

Hazlewood, A. (1951). "Optimum Pricing as Applied to Telephone Service," *Review of Economic Studies* 18 (1950-51): Reprinted in *Public Enterprise.* R. Turvey, ed. Baltimore: Penguin Books, 1968.

Hill, D.; Groves, R.; Howrey, E.P.; Lepkowski, J.; Smith, M.A.; Kline, C.; and Kohler, D. (1978). "Evaluation of Federal Energy Administration Rate Design Experiments," report prepared for the Electric Power Research Institute, Palo Alto, California (Institute for Social Research, University of Michigan, Ann Arbor, Michigan).

Houthakker, H.S. (1965). "New Evidence on Demand Elasticities," *Econometrica* 33 (April): 277-288.

Huang, S. (1975). "Building a Demand Model for Residence Phone Extensions in Minnesota," unpublished paper, Northwestern Bell Telephone Company.

Illinois Bell Telephone Company (1976). Testimony of Arnold Zellner, Illinois Docket No. 76-0409, Co. Ex. 11, July 2.

Infosino, W.J. (1976). "Estimating Flat Rate Residence Average Local Calling Rate from Demographic Variables," unpublished Bell Laboratories memorandum, August.

Irish, W.F. (1974). "A Market Analysis of the Demand for Telephone Sets by Class of Service," North Carolina Utilities Commission, December.

Johnson, N.L., and Kotz, S. (1970). *Distribution in Statistics: Continuous Univariate Distributions.* Boston: Houghton Mifflin, 1970.

Kaufman, I. (1972). "Bell Canada Message Toll Market Response System," Bell Canada, Management Science—Toll Studies, Ottawa, Ontario, October.

Kearns, T.J. (1978). "Modeling the Demand for Residence Extensions," unpublished Bell Laboratories memorandum, February.

Khadem, R. (1973a). "An Econometric Study of the Demand for Trans-Canada Long Distance Telephone Service," Bell Canada, Demand Analysis Group, Ottawa, Ontario, July.

—— (1973b). "An Econometric Demand Model of Canada-U.S. Voice Telecommunication," Computer Communications and Network Services, Bell Canada, Ottawa, Ontario, December.

—— (1974). "An Econometric Model of the Demand for Overseas Voice Telecommunications," Computer Communications and Network Services, Bell Canada, Ottawa, Ontario, October.

—— (1976). "An Econometric Forecasting Model of the Demand for International Voice Telecommunication from Canada," *Teleglobe Canada*, Montreal, October.

Konczal, E.F. (1973). "Model Building: Art and Science — Outline of Long Lines — DRI Seminar," unpublished paper, October.

Kraepelien, H.Y. (1958). "The Influence of Telephone Rates on Local Traffic," Ericsson Technics, No. 2, published by Telefonaktiebolaget, Stockholm: L.M. Ericsson.

—— (1978). "Local Telephone Pricing: The Configuration Theory," paper prepared for the Office of Telecommunications, U.S. Department of Commerce, Boulder, Colorado, March.

Krutilla, J.V. (1967). "Conservation Reconsidered," *American Economic Review* 57 (September): 777–786.

Kuh, E. (1963). *Capital Stock Growth: A Micro-Econometric Approach.* Amsterdam: North-Holland.

Kumins, L. (1967). "An Estimate of Price Elasticity of Demand for International Long-Distance Telephony," *Telecommunications Journal* 34 (June): 214–215.

Kwok, P.K.; Lee, P.C.; and Pearce, J.C. (1975). "Econometric Models of the Demand for Bell Originated Message Toll," Bell Canada, CCNS Network Market Planning, Ottawa, Ontario, January.

Lago, A.M. (1970). "Demand Forecasting Models of International Telecommunications and their Policy Implications," *Journal of Industrial Economics* 19 (November): 6–21.

Larsen, W.A.; and McCleary, S.J. (1970). "Exploratory Attempts to Model Point-to-Point Cross-Sectional Interstate Telephone Demand," unpublished Bell Laboratories memorandum, July.

Leunbach, G. (1958). "Factors Influencing the Demand for Telephone Service," paper presented at the Second International Tele-traffic Congress, The Hague, July 7–11.

Lin, W.; Hirst, E.; and Cohn, S. (1976). "Fuel Choices in the Household Sector," Oak Ridge National Laboratory, ORNL/CON-3, October.

Littlechild, S.C. (1975). "Two-Part Tariffs and Consumption Externalities," *Bell Journal of Economics* 5 (Autumn): 661–670.

—— (1977). "The Role of Consumption Externalities in the Pricing of Telephone Services," in *Pricing in Regulated Industries: Theory and Application*, J. T. Wenders, ed. Denver: Mountain States Telephone and Telegraph Co.

———, and Rousseau, J.J. (1975). "Pricing Policy of a U.S. Telephone Company," *Journal of Public Economics* 4 (February): 36-56.

Lowry, E.D. (1976). "The Demand for Telecommunications Services: A Survey," *Competition and Regulation — Some Economic Concepts.* C.F. Philips, ed. Lexington, Virginia: Washington and Lee University Press.

Maddala, G.S. (1971). "The Use of Variance Components Models in Pooling Cross-Section and Time-Series Data," *Econometrica* 39 (March): 341-358.

Mahan, G.P. (1979). *The Demand for Residential Telephone Service.* East Lansing, Michigan: Michigan State University Public Utilities Papers.

McFadden, D. (1974). "Conditional Logit Analysis of Qualitative Choice Behavior," *Frontiers of Econometrics.* P. Zarembka, ed. New York: Academic Press.

——— (1976). "The Revealed Preference of a Government Bureaucracy: Empirical Evidence," *Bell Journal of Economics* 7 (Spring): 55-72.

Michigan Bell Telephone Company (1976). Testimony, Exhibit, and Supporting Work Papers of Harold T. Shapiro, Docket U5125, September 7.

——— (1976). Rebuttal Testimony and Exhibit of E. Phillip Howrey, Docket U5125, November 22.

Mitchell B.M. (1978a). "Optimal Pricing of Local Telephone Service," *American Economic Review* 68 (September): 517-537.

——— (1978b). "Telephone Call Pricing in Europe: Localizing the Pulse," paper prepared for the Mountain Bell Economics Seminar, Keystone, Colorado, August 27-30.

Mountain Bell (1977a). Testimony of E. Milker, Montana Docket No. 6496, October 9.

——— (1977b). Testimony of J. L. Whiting, Idaho Docket No. U-1500-90 June 22.

Myskja, A., and Walmann, O. (1973). "A Statistical Study of Telephone Traffic Data with Emphasis on Subscriber Behavior," 7th International Teletraffic Congress, Stockholm, Sweden.

Naleszkiewicz, W. (1970). "International Telecommunications — Testing a Forecasting Model of Demand," *Telecommunications Journal* 37 (September): 635-638.

Nerlove, M. (1971). "Further Evidence on the Estimation of Dynamic Economic Relations from a Time Series of Cross Sections," *Econometrica* 39 (March): 359-382.

———, and Press, S.J. (1973). "Univariate and Multivariate Log-Linear and Logistic Models," Rand Corporation, December.

Nerlove, M., and Wage, S. (1964). "On the Optimality of Adaptive Forecasting," *Management Science* 10 (January): 207-224.

New England Telephone and Telegraph Company (1978). Prepared testimony of Hendrik S. Houthakker, Massachusetts Docket No. 18210—Phase II, May 15.

New Jersey Bell Telephone Company (1977). Testimony and exhibits of P.J. Dhrymes, Docket No. 7711-1136, November.

New York Telephone Company (1976a). Testimony and exhibit of S.F. Cordo, Docket No. 27100, November 17.

_____ (1976b). Testimony and exhibit of A.N. Doherty, Docket No. 27100, November 17.

_____ (1977a). Prepared testimony of T.R. Kershner, J.K. Kenney, and S. Schmidt, Docket No. 27100, July 20.

_____ (1977b). Rebuttal testimony of A.N. Doherty, Docket No. 27100, March 29.

_____ (1977c). Rebuttal testimony of S.F. Cordo, Docket No. 27100, March 29.

Northwestern Bell Telephone Company (1978a). Testimony of L. Graff, Minnesota Docket No. GR-77-1509, May 8.

_____ (1978b). Testimony of H.S. Houthakker, Minnesota Docket No. GR-77-1509, May 8.

_____ (1978c). Testimony of J.W. Johnson, Minnesota Docket No. GR-77-1509, May 8.

Ohio Bell Telephone Company (1976). Testimony and attachments to testimony of R.A. Billinghurst, Docket No. 7476-ITP-AIR, January 6.

Pacific Northwest Bell Telephone Company (1976). Testimony of L.K. Bumgarner, Docket No. U7540, Exhibits 66, 67, and 68, January 12.

Pavarini, C. (1975). "Identifying Normal and Price-Stimulated Usage Variations of Groups of Customers Part II: Individual Customer Comparison," unpublished Bell Laboratories memorandum, October.

_____ (1976a). "Residence and Business Flat Rate Local Telephone Usage," unpublished Bell Laboratories memorandum, July.

_____ (1976b). "The Effect of Flat-to-Measured Rate Conversions on the Demand for Local Telephone Usage," unpublished Bell Laboratories memorandum, September.

_____ (1979). "The Effect of Flat-to-Measured Rate Conversions on Local Telephone Usage," paper prepared for the Mountain Bell Economics Seminar, Keystone, Colorado, August 27-30, 1978, published in *Pricing in Regulated Industries*, Vol. II. J.T. Wenders, ed. Denver: Mountain States Telephone and Telegraph Co.

Perl, L.J. (1978). "Economic and Demographic Determinants of Residential Demand for Basic Telephone Service," National Economic Research Associates, Inc., March 28.

Pousette, T. (1976). "The Demand for Telephones and Telephone Services in Sweden," presented at the European Meetings of the Econometric Society, Helsinki, Finland, August.

Pyatt, G. (1972). "Some Economics of a Public Utility," Working Paper No. 28, University of Warwick, Coventry, England, September.

Ramsey, F.P. (1927). "A Contribution to the Theory of Taxation," *Economic Journal* 37 (March): 47-61.

Rash, I.M. (1971). "An Econometric Model of Demand for Residence Main Telephone Service," Bell Canada Working Paper, Ottawa, Ontario, Canada, September.

_____ (1972). "Residence Sector Demand for Message Toll Service," Bell Canada Working Paper, Ottawa, Ontario, April.

Rea, J.D., and Lage, G.M. (1978). "Estimates of Demand Elasticities for International Telecommunications Services," *Journal of Industrial Economics* 26 (June): 363-381.

Reitman, C.F. (1977). "The Measurement and Analysis of Demand Elasticities of Residential Vertical Telephone Service," unpublished Ph.D. Dissertation, New School for Social Research, May.

Rohlfs, J. (1974). "A Theory of Interdependent Demand for a Communications Service," *Bell Journal of Economics and Management Science* 5 (Spring): 16-37.

——— (1978). "Economically Efficient Bell System Pricing," unpublished Bell Laboratories memorandum, attachment H, transmitted by R.L. McGuire of AT&T to Congressman Lionel VanDeerlin, October 31.

Schankerman, M., and Stone, R. (1976). *A Project to Analyze Responses to Docket No. 20003, Final Report, Deliverable F*, prepared by T&E, Inc.

Southern Bell Telephone and Telegraph Company (1976). Testimony of R.E. Brinner, Florida Docket No. 76-0842-TP, November 29.

——— (1977). Testimony of R.E. Brinner, North Carolina Docket No. P100-Sub 45, December 6.

Southern New England Telephone Company (1977a). Curtailment testimony of J. Jeske, Connecticut Docket No. 770526, June 27.

——— (1977b). Basic Residence Revenue Curtailment Study, Connecticut Docket No. 770526, June.

——— (1977c). Study on Usage Sensitive Pricing, presented to the Public Utilities Control Authority, February 1.

Southwestern Bell Telephone Company (1978a). Testimony of B. Egan, Kansas Docket No. 110-941-U, July 7.

——— (1978b). Testimony of J.M. Griffin, Kansas Docket No. 110-941-U, July 7.

Spann, R.M. (1977). "Pricing within Urban Telephone Networks: Welfare, Allocative, and Distributional Effects of Alternative Pricing Policies," *Expanding Economic Concepts of Regulation in Health, Postal, and Telecommunications Services.* C.F. Phillips, Jr., ed. Lexington, Virginia: Washington and Lee University.

Squire, L. (1973). "Some Aspects of Optimal Pricing for Telecommunications," *Bell Journal of Economics and Management Science* 4 (Autumn): 515-525.

Stannard, R. (1978). "The Impact of Imposing Directory Assistance Charges," *Assessing the Impact of New Pricing Concepts in Public Utilities.* Harry Trebing, ed. East Lansing, Michigan: MSU Public Utilities Studies, Institute of Public Utilities, Michigan State University.

Stone, R.; Schankerman, M.; and Fenton, C. (1976). *Selective Competition in the Telephone Industry: An Independent Appraisal Based on Responses to FCC Docket 20003.* Cambridge, Massachusetts: T&E, Inc.

Stuntebeck, S. (1976). "The Relation of Day/Night Price Differentials and Message Volume Distributions for Intrastate Toll Calls," unpublished Bell Laboratories Memoradum, October.

Subissati, E. (1973). "An Econometric Model for WATS and Business Toll Services," Computer Communications and Network Services—Bell Canada, Ottawa, Ontario, August.

Swamy, P.A.V.B. (1971). *Statistical Inference in Random Coefficient Regression Models*. New York: Springer-Verlag.

Systems Applications, Inc. (1977). "The Market Outlook for Telecommunications Equipment and Services," Final Report (TF77-15), San Rafael, California, February 18.

Taylor, L.D. (1975). "The Demand for Electricity: A Survey," *Bell Journal of Economics* 6 (Spring): 74-110.

____; Blattenberger, F.R.; and Verleger, P.K. (1977). "The Residential Demand for Energy," Electric Power Research Institute, (EPRI EA-235), Palo Alto, California, January.

Theil, H., and Wage, S. (1964). "Some Observations on Adaptive Forecasting," *Management Science* 10 (January).

Thompson, H.E., and Tiao, G.C. (1971). "Analysis of Telephone Data: A Case Study of Forecasting Seasonal Time Series," *Bell Journal of Economics and Management Science* 2 (Autumn): 515-541.

Tomasek, O. (1972). "Statistical Forecasting of Telephone Time Series," *Telecommunications Journal* 39 (December): 725-731.

Von Rabenau, B., and Stahl, K. (1974). "Dynamic Aspects of Public Goods: A Further Analysis of the Telephone System," *Bell Journal of Economics and Management Science* 5 (Autumn): 651-669.

Wallace, T.D., and Hussian, A. (1969). "The Use of Error Components Models in Combining Cross-Section with Time-Series Data," *Econometrica* 55 (January): 55-60.

Wang, R.C. (1976). "Demand for Yellow Pages Advertising Space," unpublished paper, Mountain Bell, October.

Waverman, L. (1974). "Demand for Telephone Services in Great Britain, Canada and Sweden," paper presented at the Birmingham International Conference in Telecommunications Economics, Birmingham, England, May.

Weisbrod, B.A. (1964). "Collective-Consumption of Individual Consumption Goods," *Quarterly Journal of Economics* 78 (August): 471-477.

Wellenius, B. (1970). "A Method for Forecasting the Demand for Urban Residential Telephone Connections," *Telecommunications Journal* 37 (June): 262-267.

Wenders, J.T. (1976). "An Introduction to the Pricing of Telephone Service," working paper, University of Arizona, September.

Wert, G.M. (1976). "The Effects of Day/Night Price Ratios on Residence Weekday Interstate MTS Traffic," unpublished Bell Laboratories memorandum, June.

Willig, R.D. (1976). "Consumer's Surplus without Apology," *American Economic Review* 66 (September): 589-597.

Wisconsin Bell Telephone Company (1976). Prepared testimony of J. Tang, Docket No. 2U8030, Exhibit 94, July 20.

Yatrakis, P.G. (1972). "Determinants of the Demand for International Telecommunications," *Telecommunications Journal* 39 (December): 732-746.

Index

About the Author

Lester D. Taylor is a Professor of Economics at the University of Arizona. Educated at the University of Iowa (A.B.) and Harvard University (Ph.D.), he has previously taught at Harvard and at the University of Michigan, and has served on the senior staff of the President's Council of Economic Advisers and as an advisor to the governments of Colombia and Saudi-Arabia. He is a co-author with H. S. Houthakker of *Consumer Demand in the United States* and is the author or co-author of three other books and numerous articles in economics and econometrics. His research interests in recent years have focused on demand and pricing in the energy and telecommunications industries.